Lifelong Learning for Engineers and Scientists in the Information Age

Lifelong Learning for Engineers and Scientists in the Information Age

Ashok Naimpally
Fresno City College, Fresno, CA, USA

Hema Ramachandran
California State University-Long Beach, University Library
Long Beach, CA, USA

Caroline Smith
University of Nevada, Las Vegas, NV, USA

AMSTERDAM • BOSTON • HEIDELBERG • LONDON • NEW YORK • OXFORD
PARIS • SAN DIEGO • SAN FRANCISCO • SINGAPORE • SYDNEY • TOKYO

Elsevier
32 Jamestown Road, London NW1 7BY
225 Wyman Street, Waltham, MA 02451, USA

First edition 2012

British Library Cataloguing-in-Publication Data
A catalogue record for this book is available from the British Library

Library of Congress Cataloging-in-Publication Data
A catalog record for this book is available from the Library of Congress

ISBN: 978-0-12-385214-4

For information on all Elsevier publications
visit our website at elsevierdirect.com

This book has been manufactured using Print On Demand technology. Each copy is produced to
order and is limited to black ink. The online version of this book will show color figures where
appropriate.

Contents

Acknowledgments

We would like to thank the Elsevier Editorial staff Lisa Tickner, Erin Hill-Parks, Tracey Miller, and especially Sujatha Thirugnana Sambandam for their support, guidance, and in particular their patience to get this book published. We also thank the prepress production team at MPS Limited, Chennai, for their support.

Each of us would like to thank our respective institutions for providing us not only an academic home but also contact with engineering students who challenge us on a daily basis as we tackle the task of instilling in them the importance of life-long learning. This book is dedicated to all engineering educators (teaching faculty and librarians) who are continuously finding innovative ways to incorporate lifelong learning skills into the curriculum.

Dr. Naimpally would like to thank his wife Shohba. Hema Ramachandran would like to thank her husband Dr. R. Sethuraman who provided not only moral support and guidance but also held the fort at home while she tackled the most challenging project to date in her professional career. On a very personal note, it was indeed thrilling that the book was edited in the city of her birth—Chennai. The world is certainly flat!

Caroline Smith would like to thank Timothy Patrick Phillip Roberts, her darling husband and biggest supporter—"I have learned so much from you."

1 Introduction

What are necessary critical-thinking skills for engineers and scientists to have in order to enter the workforce? What skills do they need in order to stay current in a changing world throughout the course of a long career span? Educators, students, and employers alike may also ask how best to develop these skills. Certainly, in an era of prevalent electronic information, information overload, and tight academic and corporate budgets, those with the best-developed skills stand a much better chance of achieving lifelong career success. Educators seek to prepare students as thoroughly as possible for their careers, and employers seek to hire versatile and competent employees. In the United States, despite pressures of marketplace economics, academics and employers have come together into partnerships to provide students with the foundation for robust careers in engineering and science.

This book begins with the foundational elements necessary for understanding the context of lifelong learning skills as a career strength, then moves through the methods instructors and librarians use to teach and to develop students' lifelong-learning skills. Finally, we will examine the internship as a vital component to the development of a scientist or engineer with a career path open to any future direction.

Setting the education baseline, we will discuss the definition of information literacy as it is commonly understood within the university setting. Of course, accrediting boards oversee universities and must ensure universities precisely meet particular academic standards involving information literacy and lifelong-learning skills. Fortunately this is a reciprocal arrangement, as information literacy standards are embraced by all involved, including those in the Association of College and Research Libraries (ACRL) and ABET, Inc., the accrediting body involved in certifying the education of the newly minted engineer. All parties involved agree on the added value of lifelong-learning skills in the face of the information overload of the digital information age. Employers in engineering firms and scientific research centers are all equally desirous of seeing these skills developed during the educational phase of a science or engineering career. Creativity, awareness of context, and successful communication patterns are all critical elements of the successful engineer and cannot be developed without information literacy and lifelong-learning skills.

With the aid of this book, the reader will be taken through the elements necessary for creating the successful student learner. We will look at designing traditional information literacy assignments for college seniors involved in capstone courses, as

Lifelong Learning for Engineers and Scientists in the Information Age. DOI: 10.1016/B978-0-12-385214-4.00001-5

well as assignments for freshmen design classes. Literature surveys and curriculum mapping techniques will be shown as tools useful in discovering and targeting the best courses in which to incorporate information literacy. In order to ensure in-class efforts are successful, assessment tools in the context of an engineering curriculum will be discussed, including the four most powerful and popular methods used for honing and strengthening information literacy instruction.

The third element needed to create the truly information-literate scientist or engineer is the internship. This is the critical "in context" component for creating the lifelong learner. It brings to students a startling clarity of vision of their places in the information stream and with employers as they begin their careers. By looking at the history and philosophy of internships and of cooperative education, the reader will come to understand when and how to bridge the gap between academics and practice. In this context, we will present the learning contract. This is the formal assignment that binds the student, employer, and academic advisor to the standards and principles of the marketplace, the educational environment, and the formal accrediting bodies. We will provide examples of how to develop learning contracts and objectives, as well as how to assess competency methods and evaluate internship programs as a whole.

This work is intended as a guide for educators and employers—and even for students looking for a better understanding of their educational and professional future direction. Career-seeking skills are just as important in the educational process as are information-seeking skills. Fortunately for engineers and scientists, their professional support structure is fully embedded in this model.

2 Definitions of Lifelong Learning and How They Relate to the Engineering Profession

The Oxford English Dictionary (OED) defines lifelong learning thus:

> *lifelong learning n. a form of or approach to education which promotes the continuation of learning throughout adult life, esp. by making educational material and instruction available through libraries, colleges, or information technology.*
>
> *OED Online (Oxford University Press)*

The OED definition provides a good general starting point that lifelong learning is primarily conducted through educational institutions. In fact, the idea or concept of "lifelong learning" can be traced back to the ancient Greeks. At a presentation to Leningrad University, Bosco (2007) traces the classical roots of the term:

> *Although lifelong learning has become a particularly popular concept in the last several years, it is as old as human history. Lifelong learning was embodied in the works of the ancient Greeks. Plato and Aristotle described a process of learning for philosophers which extended over a lifetime. The Greek idea of a "paideia" comprised the development of a set of dispositions and capabilities which enabled and motivated the individual to continuous scholarship. Within the context of the Greek philosophers, lifelong learning was reserved for the elite social class and it was not associated with occupation or "making a living" but with the engagement in philosophic speculative inquiry.*

Bosco makes an important observation that whether we consciously recognize it or not, all of us are engaged constantly in the acquisition of information and skills as a natural process.

The definition of the term in Wikipedia provides us with more detail and distinguishes between formal and informal ways of learning and other characteristics of lifelong learning:

> *Lifelong learning is the continuous building of skills and knowledge throughout the life of an individual. It occurs through experiences encountered in the course of a lifetime. These experiences could be formal (training, counseling, tutoring, mentorship, apprenticeship, higher education, etc.) or informal (experiences, situations, etc.). Lifelong learning, also known as LLL, is the "lifelong, voluntary, and self-motivated" pursuit of knowledge for either personal or professional reasons.*

Lifelong Learning for Engineers and Scientists in the Information Age. DOI: 10.1016/B978-0-12-385214-4.00002-7

*As such, it not only enhances social inclusion, active citizenship and personal devel-
opment, but also competitiveness and employability.*

*"Lifelong Learning," Wikipedia, http://en.wikipedia.org/
wiki/Lifelong_learning, accessed September 17, 2011*

So, the concept has been in existence for some time. In this chapter, we provide a
broad overview of how various international organizations have defined lifelong
learning. Research on this topic has accelerated during the past 15–20 years as educators and policy makers worldwide have approached the task with vigor to convert
the concepts into programs. We offer a brief description of the major work in this
area to illustrate the commonalities and differences in definitions to provide our
book a frame of reference.

International Organizations

UNESCO (United Nations Educational, Scientific and Cultural Organization) was
the first organization to popularize the term in the 1960s and 1970s as a way of connecting formal and informal education. UNESCO has produced two groundbreaking reports on lifelong learning: the *Faure Report* (International Commission on
the Development of Education & Faure, 1972) and the *Delors Report* (International
Commission on Education for the Twenty-First Century, Delors, & UNESCO,
1996, 1998), both articulating the fundamental principles of lifelong learning. One
of the overarching aims of the "UNESCO Medium-Term Strategy 2008–2013,"
http://unesdoc.unesco.org/images/0014/001499/149999e.pdf, accessed September
17, 2011, for the Education Sector is "attaining quality education for all and lifelong learning." Initially, the UNESCO approach to the subject was a humanistic
one, focusing on the development of the individual with an emphasis on "learning
to learn." However, in the 1990s, UNESCO adapted its approach to lifelong learning to the needs of the "knowledge economy" and human capital development.
Despite this, the organization avoided the purely economic arguments for lifelong
learning, which is evident in its 1996 report on lifelong learning titled "Learning:
The Treasure Within." (International Commission on Education for the Twenty-first
Century et al., 1998). This report defines lifelong learning as adaptation to changes
in technology and as the continuous "process of forming whole human beings—
their knowledge and aptitudes, as well as the critical faculty and the ability to act."
UNESCO's commitment to lifelong learning is evident from its establishment in
2006 of the UNESCO Institute for Lifelong Learning (UIL) in Hamburg, Germany,
in 2006 (http://www.uil.unesco.org/, retrieved September 17, 2011). This organization is the successor to the UNESCO Institute for Education (UIE), which was
established 60 years ago. UIL's goal is to further literacy as a foundation for lifelong
learning.

The OECD (Organisation for Economic Co-operation and Development) is an
international economic organization of 34 countries founded in 1961 to stimulate
economic progress and world trade and is a forum for countries committed to

democracy and the market economy. It provides a platform for nations seeking answers to common problems by identifying good practices and coordinating the domestic and international policies of its members. In 1996, the OECD's education ministers adopted a comprehensive view of lifelong learning that covers all purposeful learning activity with the goal of "lifelong learning for all" that improves knowledge and competencies for all individuals who wish to participate in learning activities. The concept has four main features - the relevant points for our present project are summarized below:

- A *systemic view.* Viewing the demand for—and the supply of—learning opportunities as part of a connected system covering the whole life cycle and comprising all forms of formal and informal learning.
- *Centrality of the learner* shifting the focus from the supply side to the demand side of meeting learner needs.
- *Attention to learn* is recognized as an essential foundation for learning that requires developing the capacity for "learning to learn" through self-paced and self-directed learning.
- *Multiple objectives of education policy.* The life cycle view recognizes the multiple goals of education, such as personal development; knowledge development; and economic, social, and cultural objectives.

"The OECD Policy Brief on Lifelong Learning" (2004), http://www.oecd.org/dataoecd/17/11/29478789.pdf, accessed September 17, 2011, covers this topic in depth, including the following lengthy statement on the importance of lifelong learning. This statement has great relevance to the importance of and need for lifelong learning in the engineering profession:

> *A number of important socio-economic forces are pushing for the lifelong learning approach. The increased pace of globalisation and technological change, the changing nature of work and the labour market, and the ageing of populations are among the forces emphasizing the need for continuing upgrading of work and life skills throughout life. The demand is for a rising threshold of skills as well as for more frequent changes in the nature of the skills required. Firms' drive for greater flexibility has injected precariousness in jobs. There is a tendency towards shorter job tenures in the face of more volatile product markets and shorter product cycles. Career jobs are diminishing and individuals are now experiencing more frequent changes in jobs over the working life.*

The European Commission issued "A Memorandum of Lifelong Learning" in 2000, in which it recognized that the transition to a knowledge economy requires a rethinking of patterns of learning, living, and working in Europe. The definition of lifelong learning used in the Memorandum is: "all purposeful learning activity, undertaken on an ongoing basis with the aim of improving knowledge, skills, and competence and all learning activity undertaken throughout life, with the aim of improving knowledge, skills and competences within a personal, civic, social and/or employment-related perspective." The memorandum launched a Europe-wide debate on strategies for implementing lifelong learning at individual and institutional levels, and in all spheres of public and private life ("A Memorandum of Lifelong Learning," http://www.bologna-berlin2003.de/pdf/MemorandumEng.pdf, accessed September 17, 2011).

The key points of the document are the need to:

- Guarantee universal and continuing access to learning for *gaining and renewing the skills* needed for sustained participation in the knowledge society,
- Visibly *raise levels of investment* in human resources in order to place priority on Europe's most important asset—its people,
- Develop effective *teaching and learning methods* and contexts for the continuum of lifelong and lifewide learning,
- Significantly improve the ways in which learning *participation and outcomes* are *understood and appreciated*, particularly those in nonformal and informal learning,
- Ensure that everyone can easily access good quality *information and advice* about learning opportunities throughout Europe and throughout their lives, and
- Provide lifelong learning opportunities as close to learners as possible, in their own communities and supported through Information and Communications Technology Network (ICT)-based facilities wherever appropriate.

ICT refers to the information and communications technology network.

Close on the heels of this memorandum, the European Commission issued a communication entitled "Making a European Area of Lifelong Learning a Reality," http://eur-lex.europa.eu/LexUriServ/LexUriServ.do?uri=CELEX:52001DC0678: EN:NOT, accessed September 17, 2011.

Using the memorandum and the communication as a springboard, the EC has established an extensive and impressive program in lifelong learning ("Lifelong Learning Programme," http://ec.europa.eu/education/lifelong-learning-programme/doc78_en.htm, accessed September 17, 2011).

Naturally, national governments also have a vested interest in lifelong learning. It is interesting to compare the approach to lifelong learning of the United States with that of Japan. Young and Rosenberg (2006) conducted an excellent study comparing the approaches of the two countries. The paper examines the programs available to older adults in the United States and Japan and offers interesting insights for our present project. The authors conclude that there has been greater governmental acceptance of the concept in Japan whereas there has been minimal governmental intervention in the United States. In the United States the task has mostly been left to educational institutions to offer lifelong learning in the form of adult education.

In answer to the question "What is Lifelong Learning?" the Japanese Ministry of Education, Culture, Sports, Science, and Technology stated:

> In order to create an enriching and dynamic society in the 21st century, it is vital to form a lifelong learning society in which people can freely choose learning opportunities at any time during their lives and in which proper recognition is accorded to those learning achievements.
>
> Lifelong learning comprises two main aspects: the concept to comprehensively review various systems including education, in order to create a lifelong learning society; and the concept of learning at all stages of life. In other words, the concept of learning in the context of lifelong learning encompasses not only structured learning through school and social education but also learning through involvement

in such areas as sports, cultural activities, hobbies, recreation and volunteer activities. The places for conducting learning activities are also diverse, including elementary and secondary schools, universities and other institutions of higher education, citizens' public halls, libraries, museums, cultural facilities, sports facilities, lifelong learning program facilities in the private sector, companies, and offices.
The English translation is available at: "Lifelong Learning,"
http://www.qualityresearchinternational.com/glossary/
lifelonglearning.htm, accessed September 17, 2011

There are many common elements in these definitions recognizing the importance of the continuum of lifelong learning in a knowledge economy, through formal and informal channels, and through educational and vocational institutions and in the workplace. The need for purposeful self-directed learning is seen as important in all stages of a citizen's adult life to enhance the quality of one's life and to improve his or her economic standing. Many of these definitions suggest that a nation's most important asset is its people, and governments are urged to provide lifelong-learning opportunities through national programs for their citizens and to give them the ability to create their own personal pathways. In short, lifelong learning means engaging in formal and informal education on an ongoing basis, and ensuring that a person is equipped with the skills and abilities required to continue his or her own self-education beyond the end of formal schooling (Candy, 1991). Still more variants of lifelong learning include active learning, adult education, continuing education, and continuing professional development (or simply professional development).

Our book is solely concerned with lifelong learning as it relates to the education of engineering students in the United States and their subsequent performance in the workplace. The importance of the lifelong learning of engineers has firmly been established by its recognition as a program outcome in the launch of the 2000 ABET accreditation process with Criteria 3i stating that programs must demonstrate that their students attain "a recognition of the need for, and an ability to engage in lifelong learning." The evolution of ABET to an outcomes-based accreditation process is covered in detail in Chapter 3.

Recognizing the importance and significance of this topic, the National Academy of Engineering has taken up the charge by undertaking a major project: "The Lifelong Learning Imperative." The Lifelong Learning Imperative (LLI) project assesses current practices in lifelong learning for engineering professionals, reexamines the underlying assumptions, and outlines strategies for addressing unmet needs. A workshop was held in June 2009—"Lifelong Learning Project Agenda," http://llproject.org/2009-workshop-agenda/agenda, accessed September 17, 2011—to frame the project and identify critical issues for structuring the education of engineering professionals in a twenty-first-century knowledge economy. The LLI workshop initiated a national discussion on lifelong learning in the sciences and engineering and its necessity for sustaining a cutting-edge workforce in this knowledge age. A report of the findings of the workshop is published in the National Academies Press book entitled "Lifelong Learning Imperative in Engineering: Summary of a Workshop" edited by Dutta (2010). A free copy of the report is available at http://www.nap.edu/catalog.php?record_id=12866.

Some of the major themes from the workshop are summarized as follows:

- In this age of rapid technological change and knowledge creation, an engineer must continue to learn throughout his or her career.
- Learning is a continual process throughout an engineer's career.
- There is sometimes conflict between the goals of an engineer's employer and the goals of his or her current continuing education providers.
- Information technologies will play a prominent role in future lifelong learning.
- A workplace appropriately structured and augmented by access to cyberinfrastructure can be a powerful way to achieve sustainable lifelong learning.
- The engineering professional is no longer competing solely in a domestic market, but rather in a global economy.
- Further, lifelong learning should be provided through a collaborative effort among industry, academia, the government, and professional societies.

Many of these themes are also echoed in our book.

Given the half-life of a modern engineer's skills, at a minimum he or she should develop a new set of skills every decade, if not every year or so. There are relatively few lifetime jobs left in engineering, and new engineers do not even expect they will stay with one company for 25 or 30 years. Staying current on technology trends, business cycles, and inventions will demand an engineer stay up-to-date with his or her knowledge and skill set, enabling him or her to have the freedom and choice to move from job to job or from project to project, either locally or globally. Of course, this is not to forget that this global work environment is tied together by what is trending to be a global economy and a down-sized job market. Engineers working as contractors on a particular design task or programming job are not unusual. Short-term blocks of employment can often be desirable prospects to Millennials—also known as members of the Net Generation—a group linked by a revolution in communication technology and a desire to experience a variety of assignments in various locations, rather than striving for the constancy of being a company man who earns a gold watch at retirement. Innovations in communication provide exciting opportunities for personalized education, streaming to a personal data device anywhere in the world. Lifelong learning can provide the edge an engineer needs to secure the next position or the next job.

Some corporations strongly support the concept of lifelong learning. Companies such as Microsoft, Motorola, and Raytheon have their own in-house courses of study leading to certificates of expertise in various areas and at different skill levels. Companies such as these have the size, budget, and manpower to divert their engineers to lifelong learning at the expense of strict job performance and task completion. This type of lifelong learning opportunity often results in increased loyalty to the company, higher productivity, increased job satisfaction, and a rise in the quality of work. Mirroring this type of ideal lifelong learning environment is the less structured company, with a smaller organization, and fewer capital reserves. Companies such as these may not have the time, resources, or people to make any investment at all in education. They may concern themselves only with which employee was able to do a job on time and without running over budget. In either situation, it is the proactive lifelong learner who will have the knowledge base needed to secure the next position, the best jobs, and the greatest opportunities.

Lifelong learning has its foundation during the formal education that engineers receive. Lifelong learning and internships go hand in hand in creating a well-rounded engineer. Not only should a graduate emerge from the academic setting with the competency to recognize a need for lifelong learning, but he or she should have the ability to actively pursue the acquisition of knowledge. There is a certain onus that must be placed upon the newly minted engineer, and that is to "stop thinking of education as what they did for 4 years in college and come to see it as a lifetime project" (Smerdon, 1996). In order to keep their careers at pace with emerging technologies and current practices, lifelong learners should attend workshops and conferences even while they are in college so that it becomes a habit. But—at the very least— they will have to search the World Wide Web and read a book or a journal article (Mourtos, 2003) and all without the direction of a professor or mentor.

Happily, some of the best opportunities for employment start when the student engineer is still in the classroom environment and is selected for an internship opportunity with an engineering firm or corporation. An internship is a temporary apprenticeship, wherein a student receives on-the-job training to gain experience in his or her chosen field. Internships seldom last beyond 1 year, and they can be either paid or unpaid. The internship is a partnership between the student, the engineering firm providing the opportunity, and the university that provides the steady stream of eager low-cost or free workers for low-level tasks. The student gains workplace experience, familiarity with being a working engineer, and the awareness of what lifelong learning looks like on the job. The employer gets reduced cost labor and the chance to mine the next group of new engineers for talented individuals who can join their company with much training already completed. The university gains community partners for development and the goodwill of the alumni who reflect back on their university education with positive thoughts (Johnston, Taylor, & Chappel, 2001). When appraising engineering firms for suitability for internship programs, the university will look for several key factors that will contribute to students achieving a higher level of professional training. They are looking for a company with an intern mentor who can serve as a counselor, has relevant experience, has perhaps held a student internship position at some time in the past, and is someone who can serve as an advisor and guide to stimulate professional development (Guest, 2006).

As the student intern sees how life is as a professional engineer, they begin to contextualize the connection between lifelong learning and workplace literacy. Of course, each individual will differ in the amount of interest shown with regard to active lifelong learning. Even given a corporate climate with extensive learning and literacy opportunities, developing workplace literacy will still depend upon a person's background, educational experiences, and his or her professional drive (Fuller, Unwin, Felstead, Jewson, & Kakavelakis, 2007). It is known that in today's business climate, firms must be ready to deliver a "whole package" product and not simply one piece or a specific component. The new, modern engineer will have to be adept at both the technological and the commercial aspects of his or her firm and products. Streamlined staffing and corporate bottom lines require today's engineer to be skilled across multiple disciplines—able to function in a variety of job activities, rather than in specific engineering skills. Being familiar with a project's life cycle, understanding the workings of a product, and helping to market it may be part of an engineer's job, even to the point of developing close relationships with customers, leading to new product designs. The ability to operate as a systems engineer, with a grasp of all phases of a project, is an emerging professional trend in engineering (Spinks, Silburn, & Birchall, 2007). This necessitates engineers keeping up with lifelong learning and workplace literacy as—no matter how broad a professional degree program may be—it can never account for all the vagaries and practicalities one will be confronted with on the job.

And so, with foundational support from the educators instructing the upcoming generations of engineers, from the accrediting organization ABET, and from members of the engineering profession, themselves, lifelong learning is becoming a way of life for those who choose a career in engineering.

References

Bosco, J. (June 2007). *Lifelong learning: what? why? how?* Retrieved August 2011, from <http://homepages.wmich.edu/~bosco/docs/LifelongLearning-2.pdf>.

Candy, P. C. (1991). *Self-direction for lifelong learning: A comprehensive guide to theory and practice* (1st ed.). San Francisco, CA: Jossey-Bass.

Dutta, D. (Ed.). (2010). *Lifelong learning imperative in engineering: Summary of a workshop.* Washington, DC: National Academies Press.

Fuller, A., Unwin, L., Felstead, A., Jewson, N., & Kakavelakis, K. (2007). Creating and using knowledge: An analysis of the differentiated nature of workplace learning environments. *British Educational Research Journal, 33*, 743. 10.1080/03043790600644396.

Guest, G. (2006). Lifelong learning for engineers: A global perspective. *European Journal of Engineering Education, 31*, 273. 10.1080/03043790600644396.

International Commission on Education for the Twenty-First Century, Delors, J. & UNESCO. *Learning: The treasure within: Report to UNESCO of the international commission on education for the twenty-first century* (2nd (pocketbook) ed.). Paris: UNESCO.

International Commission on the Development of Education & Faure, E. (Eds.). (1972). *Learning to be; the world of education today and tomorrow [by] Edgar Faure [and others].* Paris: UNESCO.

Johnston, S., Taylor, E. & Chappel, A. (2001). *UTS engineering internships: A model for active work place learning.* Retrieved August 13, 2011, from <http://citeseerx.ist.psu.edu/viewdoc/summary?doi=10.1.1.58.1044>.

Mourtos, N. J. (2003). Defining, teaching and assessing lifelong learning skills. In: *Proceedings of the 33rd ASEE/IEEE Frontiers in Education Conference, November 5–8, 2003, Boulder, CO* (pp. T3B14–T3B19).

Oxford University Press. Lifelong learning. In: *OED online.* Retrieved from <http://www.oed.com/view/Entry/108114?redirectedFrom=lifelong%20learning%20>.

Smerdon, E. J. (1996). Lifelong learning for engineers: Riding the whirlwind. *The Bridge, 26*(1–2), May 24, 2010.

Spinks, N., Silburn, N. L. J., & Birchall, D. W. (2007). Making it all work: The engineering graduate of the future, a UK perspective. *European Journal of Engineering Education, 32*, 325–335. 10.1080/03043790701278573.

Young, K., & Rosenberg, E. (2006). Lifelong learning in the United States and Japan. *The LLI [Lifelong Learning Institute] Review, 1*(1), 69–85. Retrieved from <http://usm.maine.edu/olli/national/lli-review.jsp>.

3 Accreditation of Engineering Programs and Their Relationship to Lifelong Learning

Accreditation of Engineering Programs in the United States

In the United States, professional certification for the engineering field—mostly through university accreditation—falls under the aegis of ABET, Inc. Accreditation is the process by which competency and authority are established in a particular métier, and for American engineers and engineering schools, ABET is the authoritative body making these determinations. However, ABET does not act alone, and—in fact—ABET is part of an engineering leadership internationally. Through cooperation and mutual recognition, accrediting organizations work to keep engineering programs up to par in order to help foster the discipline and create a marketplace of skilled individuals that establish the engineering standards so critical to professional engineering.

ABET and its predecessors have a long history, and they have evolved together for the betterment of the profession. The Engineers' Council for Professional Development (ECPD) was established in 1932 in the wake of a survey conducted by the major professional engineering societies that revealed that joint organization was necessary in order to bolster the engineering profession. The ECPD mainly focused on the following aspects:

- Guidance: Supplying information to current and potential engineering students.
- Training: Developing plans for personal and professional development.
- Education: Appraising engineering curricula and maintaining a list of accredited curricula.
- Recognition: Developing methods whereby individuals could achieve recognition by the profession and the general public.

The seven engineering societies that came together and founded the ECPD organization were: the American Society of Civil Engineers (ASCE), the American Institute of Mining and Metallurgical Engineers (now the American Institute of Mining, Metallurgical and Petroleum Engineers (AIME)), the American Society of Mechanical Engineers (ASME), the American Institute of Electrical Engineers (now the Institute of Electrical and Electronics Engineers (IEEE)), the Society for the Promotion of Engineering Education (now the American Society for Engineering Education—ASEE), the American Institute of Chemical Engineers (AIChE), and the National Council of State Boards of Engineering Examiners (now the National Council of Examiners for Engineering and Surveying—NCEES).

Lifelong Learning for Engineers and Scientists in the Information Age. DOI: 10.1016/B978-0-12-385214-4.00003-9

In 1936, only 4 years after its establishment, ECPD evaluated and accredited its first engineering degree program, and 10 years later it evaluated its first engineering technology program. Thus, the seeds were laid for the extensive program that is in place today. By 1947, ECPD had accredited 580 undergraduate programs at 133 institutions of higher learning.

ECPD was renamed the Accreditation Board for Engineering and Technology (ABET) in 1980: a name change to reflect its main mission, the accreditation of engineering curricula in higher education institutions. In 2005, the organization formally changed its name from the "Accreditation Board for Engineering and Technology" to ABET, Inc. By the late 1980s, the organization also became a consultant to both new and established international accreditation boards. Presently, ABET accredits some 2,900 programs at more than 600 colleges and universities nationwide. ABET now has 30 member societies that volunteer their services as evaluators and provide leadership and quality assurance in applied science, computing, engineering, and technology education, in addition to serving on the ABET Board (see http://abet.org/gov.shtml). In 1997, ABET was recognized by the Council for Higher Education Accreditation as the authoritative voice of engineering education (see http://www.chea.org/).

For the first part of its history, ABET's accreditation criteria concentrated almost entirely on curricula and the number and quality of faculty and facilities. However, a paradigm shift in thinking began in the mid-1990s as the engineering community began a series of in-depth discussions on the nature and scope of accreditation requirements. The accreditation process was deemed to be inflexible and was accused of not allowing or encouraging innovation at academic institutions. This rigidity was keeping engineering programs out of touch with the true future needs of professional engineers. The development of Engineering Criteria 2000 (EC2000) was the result of this intense national debate. In five short years, ABET moved from discussion to the actual implementation of EC2000.

Key Milestones for Implementation of EC2000

Program Criteria published for 1 year comment (see http://www.abet.org)	December 1996
EAC reviews Pilot Study	July 1997
EAC reviews/revises General Criteria based on comments	July 1997
EAC reviews/revises Program Criteria based on comments	July 1997
Pilot Study, additional schools visited	September to December 1997
ABET Board votes on Revised General/ Program Criteria	November 1997
Criteria published, effective for 1998/1999 visit cycle	December 1997
EAC reviews Pilot Study	July 1998
Begin 3-year phased implementation	Fall 1998 visit cycle
Full implementation of Criteria 2000	Fall 2001

Source: "Engineering Criteria 2000" (3rd ed.). http://www.ele.uri.edu/faculty/daly/criteria.2000.html, accessed September 16, 2011.

EC2000

Launched in late 1997 following a pilot study, EC2000 was a revolutionary new way of thinking about teaching and learning. Instead of focusing on teaching (inputs)— what is taught and how—the focus is on what students learn (outputs). Also, EC2000 stresses the fundamental concept of continuous program improvements while taking into account the missions and goals of academic institutions and programs. In effect, ABET had moved to an "outcomes"-based method of assessment. This shift in thinking mirrors current pedagogical thinking and brings ABET in line with academic standards that require a fixed set of student learning outcomes. Explicit learning statements define the knowledge, skills, and abilities expected of each student at the end of a particular course or course of study. These student learning outcomes are assessed at the end of each course and at the end of the entire course of academic study. Though a set of prescribed expectations may sound restrictive, it actually allows institutions to be creative in developing innovative programs that reflect the needs of future engineers who will be working in a fast-paced and changing global economy. In short, the students all arrive at the point of being educated engineers, but how they get there is a creative and dynamic journey that can vary from instructor to instructor or from university to university. In the decade since the implementation of EC2000, we are beginning to see revised and revamped curricula and new curricula as faculty and administrators experiment and design new programs to honor the spirit of EC2000 on their respective campuses.

ABET Program Outcomes

At the center of EC2000 are the ABET Program Outcomes (for baccalaureate programs)—often referred to as Criterion 3a–k—reproduced in the box titled "2010–2011 Criteria for Accrediting Engineering Programs" from the ABET website: http://www.abet.org/Linked%20Documents-UPDATE/Criteria%20and%20PP/E001%2010-11%20EAC%20Criteria%2011-03-09.pdf, accessed September 3, 2011.

Program results are outcomes (a) through (k), plus any additional abilities that might be articulated by the program. Program outcomes must foster attainment of program educational objectives. There must be an assessment and evaluation process that periodically documents and demonstrates the degree to which the program outcomes are attained.

These outcomes can be divided into two broad categories: those that are concerned with "hard" or technical skills and those that address "soft" skills. Criteria 3a–c, e, and k are concerned with the technical foundations of engineering (hard skills). Before EC2000, university engineering programs paid limited attention to writing and other communication skills. However, during the intense national debate prior to the launch of EC2000, it became clear that programs should give much more prominence to the soft skills, such as writing; communicating; and understanding ethical, global, and contemporary issues, along with the concept of lifelong learning. The professional

---

2010–2011 Criteria for Accrediting Engineering Programs

Criterion 3. Program Outcomes

Engineering programs must demonstrate that their students attain the following outcomes:

a. an ability to apply knowledge of mathematics, science, and engineering
b. an ability to design and conduct experiments, as well as to analyze and interpret data
c. an ability to design a system, component, or process to meet desired needs within realistic constraints such as economic, environmental, social, political, ethical, health and safety, manufacturability, and sustainability
d. an ability to function on multidisciplinary teams
e. an ability to identify, formulate, and solve engineering problems
f. an understanding of professional and ethical responsibility
g. an ability to communicate effectively
h. the broad education necessary to understand the impact of engineering solutions in a global, economic, environmental, and societal context
i. a recognition of the need for, and an ability to engage in lifelong learning
j. a knowledge of contemporary issues
k. an ability to use the techniques, skills, and modern engineering tools necessary for engineering practice.

---

literature abounds with articles about the definition of soft skills and how to teach them as part of the engineering curriculum. Two well-cited articles are worth mentioning. The first, by Moore and Voltmer (2003), urges the profession to return to its original mission—that of a service profession—the essence of which is service, professionalism, creativity, and problem solving. The other article, by Ragsdell (2000), is a sociological study reporting the empirical findings from a series of action research projects conducted in several engineering companies over a period of 3 years. Each company was implementing a new company-wide strategy to make them more competitive. One of the major conclusions of this study is that senior managers felt that their engineers had the hard skills to comply with the new strategy, but lacked the soft—or people—skills to cope with the cultural or organizational change. It is evident that future engineers must not only have a thorough understanding of engineering principles, but must also be skilled in communicating their knowledge, must work well with people at all levels, and must adapt well to organizational, societal, and cultural change. Engineers—working in teams—must act ethically and responsibly in an increasingly global world. Criteria 3d and f–j address these important soft skills—abilities that are complementary and essential companions to technical skills.

ABET Criteria 3i

This book is concerned in particular with the ABET Criteria 3i for lifelong learning—"a recognition of the need for, and an ability to engage in lifelong learning"—and strategies for its implementation. (Please note that during the 2009–2010 program cycle, Criteria 3i—related to lifelong learning—was moved up to 3h but has since been moved back to 3i in the current cycle.) Due to its very nature, lifelong learning (ABET Criteria 3i) occurs during the entire professional life of a scientist or engineer—a period that is anywhere up to 40 years long—while the formal degree-based education of the professional can range from 4 to 10 years, a fraction of the professional period. Due to rapid changes in technology, and in economic and sociocultural systems on this planet, the process of "introducing" lifelong learning can include:

- Mastering the techniques—at least the ones available during the years of formal study—of lifelong learning,
- Creating a strong awareness of lifelong learning, and
- Developing the instinct to incorporate lifelong learning into one's life and routine during the rough-and-tumble of the professional's working world.

Mastering the Techniques of Lifelong Learning

Mastering the techniques of lifelong learning first involves increasing the independence of the student in the learning process. Litzinger, Wise, Lee, Simpson, and Joshi (2001), summarizing the extensive work of Candy (1991), give three broad strategies for increasing independent learning:

- Offering courses focused on developing skills important to self-directed learning; some examples of this are classes in information literacy, self-management, and critical thinking.
- Giving students an opportunity to practice those skills.
- Giving students an opportunity to develop those skills and engage in self-reflection.

The Foundation Coalition (FC), an important resource in engineering education (composed of a group of eight engineering coalitions and funded by the National Science Foundation), provides significant resources and monitors research on the implementation of ABET Criteria 3i: http://www.foundationcoalition.org/home/keycomponents/assessment_eval/outcome_i.html.

Summarizing the research referenced on the Foundation Coalition's Criteria 3i-related webpage, there are many examples of assignments that would work well in learning techniques useful for lifelong learning:

1. Requiring a literature review.
2. Involving attendance at meetings of professional societies.
3. Acquiring sources for continuing education in a given area of the discipline for varying assignments, including for a literature review—for instance using SciFinder Scholar or Compendex or other databases that provide access to peer-reviewed literature; using the Internet to search for non-peer-reviewed literature; taking formal short courses; attending workshops and webinars; interviewing and networking with practicing engineers or scientists; reading and understanding trade literature related to the area under study; this will enable the student to explore many different research methods.

4. Giving writing assignments that require explanations of the importance of lifelong learning in a given area of study.

Scientists and engineers learn at several broad levels:

- Understanding existing theories and concepts based on the past work in the chosen field of science or engineering,
- Gaining skills in problem solving by using instruments and equipment, and learning laboratory techniques,
- Synthesizing and obtaining more strategic, higher levels of understanding from learning in seemingly disparate areas. Examples are concepts [of ...], attending lectures by working scientists, reading published papers, and watching webinars,
- Being "creative" in the chosen field of study and to formulate thinking outside the box solutions to problems, and
- Working in a team.

For a professional to keep up-to-date in a competitive world in which younger professionals enter the workforce involves the ability of the professional to keep pace in all five areas listed above. The first ability involves expanding one's knowledge base during one's lifetime, and the second ability involves learning to operate newer pieces of equipment and instruments. These two skills actually form the core of undergraduate work. The third and fifth abilities can be developed through assignments that are multidisciplinary in nature and/or are carried out in a team setting.

Awareness of Lifelong Learning

Awareness of the need for lifelong learning is likely to be less of a problem for the current generation of professionals working in a communicative, wired, globalized environment. However, "information overload" presents other challenges, such as the need to develop higher critical-thinking skills to determine (for instance) the authoritativeness and appropriateness of different types of information. The awareness that may need to be developed is that much of the learning that will prove useful to the professional is likely to come from traditional sources. This strong awareness could best be acquired by giving related assignments within courses (items 1 through 4 from the Foundation Coalition's research).

The NAE (National Academy of Engineering) newsletter "The Bridge" offers one of the best descriptions of the challenges for engineers attempting to keep up in their disciplines in the article by Smerdon (1996) aptly entitled "Lifelong Learning for Engineers: Riding the Whirlwind":

> *A decade ago, a group of experts estimated the half-life of an engineer's technical skills—how long it would take for half of everything an engineer knew about his or her field to become obsolete. For mechanical engineers it was 7.5 years. For electrical engineers it was 5. And for software engineers, it was a mere 2.5 years, less time than it takes to get an undergraduate degree. Today, those numbers are surely even smaller.*

and

> *Think about it. In some specialties, engineers must update half of everything they know every couple of years, all the while working full-time to design products according to the best standards of the moment—which might change next month. In even the slower-paced fields, engineers must reinvent themselves at least once a decade.*
>
> *Add to this already overwhelming situation the fact that engineers can no longer expect to spend their entire working lives specializing in a niche area. As technology changes and production life cycles ebb and flow, engineers must be ready to switch jobs—not just between companies, but even within a company—and must "be prepared to switch nimbly to a new field when the old one peters out."*

Paton (2002) suggests that engineers should expect to spend from 100 to 300 hours on continuing education each year. He also tells us that the rate of change in areas such as information technology, semiconductors, telecommunications, materials science, and chemical technology is approaching 25% per year. Another interesting observation by Paton is that once a company has developed and implemented a new technology, it will move on to a newer technology. Engineers unable to educate themselves will not be selected for the new project and will eventually be moved into lower-tier roles.

Given that the half-life of engineering knowledge is anywhere from 2.5 to 7.5 years, depending on the discipline, students need to be provided with the relevant tools and techniques to keep abreast of developments in their respective engineering discipline, in addition to related areas such as business. ABET Criteria 3i on lifelong learning is designed to address this issue, and yet—without a doubt—it is one of the most challenging of the 11 outcomes to assess. How do we know or how can we ever know that we have succeeded in meeting this outcome? The most conclusive way to measure this outcome would be to interview and survey students after they graduate to ascertain if they are pursuing lifelong-learning goals.

Marra, Camplese, and Litzinger's article (1999) is one of the first post-EC2000 publications to articulate the issues surrounding the lifelong-learning outcome. Their article summarizes the results of a preliminary literature review of lifelong learning and engineering education and discusses plans for assessing the lifelong learning of Penn State students, including some data from a survey of their recent graduates. As Litzinger and Marra (2000) rightly observe, previously it was enough for academic institutions to offer courses in lifelong learning for practicing engineers as part of their program, but they are now required to demonstrate they have met this outcome for their current students. Litzinger and Marra ask what the critical skills and attributes of lifelong learning are and how we can adapt our curricula to meet these demands. They also discuss some attributes and characteristics of lifelong learners and ways to assess them.

Mourtos (2003) further develops these concepts and ideas using Bloom's Taxonomy and suggests that "a recognition of the need for" requires skills in the affective domain and the "ability to engage" requires skills in the cognitive domain. Mourtos offers five learning objectives, one for each level of the affective domain:

1. Willingness to learn new material on their own,
2. Reflecting on their learning process,
3. Participation in professional societies' activities,
4. Reading engineering articles and books outside of class, and
5. Attending extracurricular training or planning to attend graduate school.

He places the second outcome element in the more familiar cognitive domain and offers nine learning objectives:

1. Observe engineering artifacts carefully and critically, to reach an understanding of the reasons behind their design,
2. Access information effectively and efficiently from a variety of sources,
3. Read critically and assess the quality of information available: question the validity of information, including that from textbooks or teachers,
4. Categorize and classify information,
5. Analyze new content by breaking it down, asking key questions, comparing and contrasting, recognizing patterns, and interpreting information,
6. Synthesize new concepts by making connections, transferring prior knowledge, and generalizing,
7. Model by estimating and simplifying, and by making assumptions and approximations,
8. Visualize: create pictures in your mind that help you "see" what the words in a book describe, and
9. Reason by predicting, inferring, using inductions, questioning assumptions, using lateral thinking, and inquiring.

Mourtos makes the case succinctly for the need for information literacy as a vital tool for lifelong learning:

> *Assuming that the graduate school option has been exhausted, engineers can stay current throughout their career by attending short courses, workshops, seminars, and conferences in their own as well as in new, emerging fields. However, it is not practical to expect that all the new knowledge we will ever need at some point in our careers can be acquired through these venues. Sooner or later, one will have to search the worldwide web, go to the library or the bookstore, and eventually sit down with a book, an article or some other reference to learn on his/her own. It is in this context that lifelong learning skills need to be defined, taught, and practiced.*

ABET formed out of a discussion among professional engineers. Those engineers and academics of the 1930s saw a need for change and took action to codify the engineering profession as well as the classroom skills and tools needed to train engineers. Decades of academic and technical progress (almost a "Moore's law" in engineering technology and influence) brought to the fore the need to rethink the educational and accrediting partnership. Increased communication and the globalization of the work environment created a need for a new type of engineer: an adaptive individual, capable of nimble thinking, self-education, cooperation, and technical skill. The type of learning imparted in the university classroom is nurtured as a lifelong skill in the professional realm under the guidance of ABET, Inc. Educating the engineers of the future, and keeping them engaged and professionally active is the essential task of the ABET accreditation process.

References

Candy, P. C. (1991). *Self-direction for lifelong learning: A comprehensive guide to theory and practice* (1st ed.). San Francisco, CA: Jossey-Bass.

Litzinger, T. A., & Marra, R. M. (2000). Lifelong learning: Implications for curricular change and assessment. In: *2000 ASEE Annual Conference and Exposition: Engineering Education Beyond the Millennium, June 18–21, 2000* (pp. 4083–4090).

Litzinger, T., Wise, J., Lee, S., Simpson, T., & Joshi, S. (2001). Assessing readiness for lifelong learning. Paper presented at the *American Society for Engineering Education Annual Conference & Exposition*. Retrieved from <http://openedpractices.org/files/Engineering%20and%20LLL%20at%20Penn%20St..pdf>.

Marra, R. M., Camplese, K. Z., & Litzinger, T. A. (1999). Lifelong learning: A preliminary look at the literature in view of EC2000. *Proceedings—Frontiers in Education Conference.*

Moore, D. J., & Voltmer, D. R. (2003). Curriculum for an engineering renaissance. *IEEE Transactions on Education, 46*(4), 452–455. 10.1109/TE.2003.818754.

Mourtos, N. J. (2003). Defining, teaching and assessing lifelong learning skills. In: *Proceedings of the 33rd ASEE/IEEE Frontiers in Education Conference, November 5–8, 2003, Boulder, CO* (pp. T3B14–T3B19).

Paton, A. E. (2002). What industry needs from universities for engineering continuing education. *IEEE Transactions on Education, 45*(1), 7–9.

Ragsdell, G. (2000). Engineering a paradigm shift? An holistic approach to organisational change management. *Journal of Organizational Change Management, 13*(2), 104–120.

Smerdon, E. J. (1996). Lifelong learning for engineers: Riding the whirlwind. *The Bridge, 26*(1/2) , May 24, 2010.

4 Information Literacy and Lifelong Learning

The Association of College and Research Libraries (ACRL) has developed an in-depth explication of information literacy (IL). Since ACRL is the academic arm of the American Library Association (ALA), this philosophy and educational practice of IL influences the lifelong learning of engineers via the selection of research and academic materials by librarians and via the course content supplied by librarians in their instructional roles.

In the broadest sense, ACRL defines IL as

> *Information literacy is a set of abilities requiring individuals to "recognize when information is needed and have the ability to locate, evaluate, and use effectively the needed information." Information literacy also is increasingly important in the contemporary environment of rapid technological change and proliferating information resources. Because of the escalating complexity of this environment, individuals are faced with diverse, abundant information choices—in their academic studies, in the workplace, and in their personal lives. Information is available through libraries, community resources, special interest organizations, media, and the Internet and increasingly, information comes to individuals in unfiltered formats, raising questions about its authenticity, validity, and reliability. In addition, information is available through multiple media, including graphical, aural, and textual, and these pose new challenges for individuals in evaluating and understanding it. The uncertain quality and expanding quantity of information pose large challenges for society. The sheer abundance of information will not in itself create a more informed citizenry without a complementary cluster of abilities necessary to use information effectively.*
>
> *Information literacy forms the basis for lifelong learning. It is common to all disciplines, to all learning environments, and to all levels of education. It enables learners to master content and extend their investigations, become more self-directed, and assume greater control over their own learning. An information-literate individual is able to:*
>
> - *Determine the extent of information needed*
> - *Access the needed information effectively and efficiently*
> - *Evaluate information and its sources critically*
> - *Incorporate selected information into one's knowledge base*
> - *Use information effectively to accomplish a specific purpose*
> - *Understand the economic, legal, and social issues surrounding the use of information, and access and use information ethically and legally*
>
> *From http://www.ala.org/ala/mgrps/divs/acrl/standards/*
> *informationliteracycompetency.cfm.*

Lifelong Learning for Engineers and Scientists in the Information Age. DOI: 10.1016/B978-0-12-385214-4.00004-0

ACRL then takes this broad standard and refines it specifically for science and engineering library professionals, creating a uniform standard of IL for them:

> *Information literacy in science, engineering, and technology disciplines is defined as a set of abilities to identify the need for information, procure the information, evaluate the information and subsequently revise the strategy for obtaining the information, to use the information and to use it in an ethical and legal manner, and to engage in lifelong learning. Information literacy competency is highly important for students in science and engineering/technology disciplines who must access a wide variety of information sources and formats that carry the body of knowledge in their fields. These disciplines are rapidly changing and it is vital to the practicing scientist and engineer that they know how to keep up with new developments and new sources of experimental/research data.*
>
> *Science, engineering, and technology disciplines pose unique challenges in identifying, evaluating, acquiring and using information. Peer reviewed articles are generally published in more costly journals and, therefore, not always available. Gray literature requires knowledge of the agency/organization publishing the information. Much of science, engineering and technology is now interdisciplinary and, therefore, requires knowledge of information resources in more than one discipline. Information can be in various formats (e.g. multimedia, database, website, data set, patent, Geographic Information System, 3-D technology, open file report, audio/visual, book, graph, map) and, therefore, may often require manipulation and a working knowledge of specialized software.*
>
> *Science, engineering, and technology disciplines require that students demonstrate competency not only in written assignments and research papers but also in unique areas such as experimentation, laboratory research, and mechanical drawing. Our objective is to provide a set of standards that can be used by science and engineering/technology educators, in the context of their institution's mission, to help guide their information literacy-related instruction and to assess student progress. The field of mathematics is not included in the standards.*
>
> *Based on the ACRL Information Literacy Competency Standards for Higher Education, five standards and twenty-five performance indicators were developed for information literacy in Science & Engineering/Technology. Each performance indicator is accompanied by one or more outcomes for assessing the progress toward information literacy of students of science and engineering or technology at all levels of higher education.*
>
> ALA/ACRL/STS Task Force on Information Literacy for
Science and Technology

Comparison of ACRL and ABET Standards

Since librarians are instruction partners in the education of engineers and scientists at both the graduate and the undergraduate levels, it is no surprise that ACRL's standards for science, engineering, and technology would match ABET's standards so well. The shared vision and common purpose of the two organizations is easily seen in following chart:

ABET Accreditation Criteria Mapped to ACRL IL Standards

ABET Criterion	ACRL Standards for Science and Engineering/ Technology Performance Indicators
a. An appropriate mastery of the knowledge, techniques, skills, and modern tools of their disciplines	3.1.a; 3.3.c; 3.6.a; 3.6.b; 4.1.b; 4.1.d; 4.2.a; 4.3.b; 2.1.a; 1.3.a–f
b. An ability to apply current knowledge and adapt to emerging applications of mathematics, science, engineering, and technology	3.1.b; 3.1.c; 3.2.a; 3.4.a–g; 3.6.c; 3.7.a; 4.1.a; 4.1.b; 4.2.a; 1.4.a–e; 5.1.a–d
c. An ability to conduct, analyze, and interpret experiments, and apply experimental results to improve processes	3.2.b; 3.3.a–b; 3.5.b; 3.7.b; 4.1.a–c; 4.2.a–b; 4.3.c; 2.1.a; 1.2.d; 3.2.e; 4.5.a–c
d. An ability to apply creativity in the design of systems, components, or processes appropriate to program educational objectives	4.2.b; 4.4.a–b
e. An ability to function effectively on teams	3.6.c; 2.3.d; 3.5.b
f. An ability to identify, analyze, and solve technical problems	1.1.b
g. An ability to communicate effectively	3.6.c; 4.1.b; 4.3.a–d; 2.3.d; 1.2.b; 3.5.a; 3.5.c; 4.3.a–c; 4.6
h. A recognition of the need for, and an ability to engage in lifelong learning	3.1.a–3.7.b; 4.1.a–4.6.d; 2.2.a–f; 3.2.b–d; 3.2.f–g
i. An ability to understand professional, ethical, and social responsibilities	3.2.a–b; 4.1.a–d; 4.2.a–g
j. A respect for diversity and a knowledge of contemporary professional, societal, and global issues	3.2.d
k. A commitment to quality, timeliness, and continuous improvement	3.2.a; 3.5.a; 3.7.c

Please note that at the time Sapp Nelson and Fosmire (2010) wrote this article, the lifelong-learning criteria was moved to 3h, but it has subsequently been moved back to 3i for the 2010–2011 program cycle. From http://www.abet.org/Linked%20Documents-UPDATE/Criteria%20and%20PP/E001%2010-11%20EAC%20 Criteria%2011-03-09.pdf.

The relationship between IL and lifelong learning is now inexorably joined to improvements and cohesion in academics. Librarians and instructors have documented a common and supporting purpose when it comes to educating science and technology students. But developments at the professional level need to be successfully merged into the classroom—for a purposefully inculcated foundation of IL is the basis for any set of lifelong-learning skills. It is impossible to have that ultimate professional skill set without instruction and practice in research, discernment, and the articulation of ideas. These types of critical-thinking skills can be taught and learned through guidance and practice in the classroom. Insights and direction from practiced academic professionals make developing IL skills a much easier process than if one had to develop these skills alone and at point-of-need on the job. The value of the ability to know where to look for information—and then how to analyze, synthesize, and distribute it—cannot be overstated. There has been a global move to a knowledge-based economy, but it is not clear whether or not businesses and corporations are addressing IL gaps with professional development as a common practice. The world of business and industry, by virtue of the digital age itself, has an investment in knowledge management as an essential aspect of its functionality. Corporate leaders are aware of the value of employees with IL skills as a "new economy" skill set, and without this skill many workers might be ill-equipped to deal with the information overload that is common in the science and engineering professions (O'Sullivan, 2002).

The contribution of IL to lifelong learning and the career of engineers and scientists can be seen in the value added to one's skill set and the resulting demand from employers. These skills "should be invaluable for all design projects" (Clarke & Coyle, 2011), and they can be leveraged "for product development research throughout their careers" (Clarke & Coyle, 2011). Most assuredly, engineers and scientists will be expected to write much more in their careers than they likely expected as students or novice professionals first entering the workforce. As they move ahead professionally, they will find increased demand upon them to produce written materials, such as white papers, grant applications, and requests for proposals, and their IL and lifelong-learning skills will serve them well.

In order to ensure that students' information development needs will be met, librarians and instructors must continue to work together to develop curricula and tutorials aimed at enhancing the student learning experience. Continuing education for teaching professionals is just as important as it is for those in technology-based jobs. The ACRL maintains an instruction wiki specifically for those engaged in engineering education who are concerned with ABET standards and academic curricula. Referencing http://wikis.ala.org/acrl/index.php/Information_Literacy_in_ Engineering, one can access a well-maintained and well-reviewed list of articles and curricula related to developing lifelong-learning skills, thus enabling one to perform an easy, baseline literature review on this topic.

References

ALA/ACRL/STS Task Force on Information Literacy for Science and Technology. *Information literacy standards for science and Engineering/Technology.* Retrieved September 12, 2011, from <http://www.ala.org/ala/mgrps/divs/acrl/standards/ infolitscitech.cfm>.

Clarke, J. B., & Coyle, J. R. (2011). A capstone wiki knowledge base: A case study of an online tool designed to promote life-long learning through engineering literature research. *Issues in Science and Technology Librarianship, 65*(Spring).

O'Sullivan, C. (2002). Is information literacy relevant in the real world? *Reference Services Review, 30*(1), 7.

Sapp Nelson, M., & Fosmire, M. (2010). *Engineering librarian participation in technology curricular redesign: Lifelong learning, information literacy, and ABET criterion 3.* Retrieved September 12, 2011, from <http://soa.asee.org/paper/conference/paper-view. cfm?id=23424>

5 Creativity in Engineering, Information Literacy, and Communication Patterns of Engineering

Nature of Engineering

It is important for librarians to understand the nature of engineering so that they can collaborate more effectively with teaching faculty and can design appropriate assignments in information literacy (IL). The nature and essence of engineering is often not understood correctly by non-engineers. The usual image of an engineer is of someone who is deeply involved in the technical aspects of a project and only concerned with minutiae, and who is walking around in a hard hat measuring different parts of a project with blueprints and charts. And this is, of course, true when one interacts with an engineer (for instance) on a construction site. However, at the very heart of engineering are notions of creativity, innovation, problem solving, and design. It is these aspects of the discipline that attract students to engineering and keep working engineers engaged and enthusiastic about their profession.

Engineers have played a major role in product design and development for centuries. All of the products we use every day were designed and developed by someone with an engineering mind-set whether they were a certified engineer, an independent inventor, or a part of a company or team. Creative engineers are the ones who solve the world's technological problems and improve the quality of our everyday lives. Charyton and Merrill (2009) provide an excellent history of research on creativity and its application to engineering as a prelude to a report on the development of a tool to assess the creativity of engineering undergraduate students. Some salient parts of their review are presented in this chapter to deepen our understanding of creativity as it applies to engineering. The essence of creativity can be defined as a preference for thinking in novel ways and the ability to produce work that is novel and appropriate (Sternberg, 1999; Weisberg, 1986). Felder (1987), one of the leading researchers in engineering education, goes so far as to say that it is the profession's responsibility to produce creative engineers "or at least not to extinguish the creative spark in our students." So, it is necessary to provide opportunities for our students to exercise and develop their creative skills. Elliott (2001) went further and stated: "I think that if engineers are not creative, they are not engineers."

Simonton (2000) states that "Creativity is certainly among the most pervasive of all human activities. Homes and offices are filled with furniture, appliances, and

Lifelong Learning for Engineers and Scientists in the Information Age. DOI: 10.1016/B978-0-12-385214-4.00005-2

other conveniences that are products of human inventiveness." Highly creative people redefine problems, analyze ideas, persuade others, and take reasonable risks in order to generate ideas (Sternberg & Dess, 2001).

But how is creative thinking related to IL? Goad (2002), in his book "Information Literacy and Workplace Performance," makes an excellent connection between creativity, innovation, and risk taking, and then shows how they relate to IL. Chapter 5 of his book could easily be applied to any engineering workplace where one has to "innovate or evaporate." The following lengthy quotation offers us the best explanation of the connection between creativity and IL:

> *On the surface, it might appear that creative thinking has little to do with being information literate. After all, isn't information literacy the ability to acquire and use information? It's achieved through a step-by-step, well organized process, and creativity is anything but. Information literacy is that, but much more, and the key is thought. Thinking, as discussed in the previous chapter is required at all steps along the way from information need to its application for problem-solving and decision making. ... Continuously coming up with great ideas may well dictate the information needed. Being creative when, say, you're considering alternative solutions to a problem (the alternatives resulting from an information search), may produce far better results than taking a more traditional thinking approach.*
>
> *Goad (2002, p. 84)*

The Design Process

Engineers express their creativity through the design process to solve problems within constraints. The following slide by Salomon Davila of Pasadena City College [part of a presentation at the American Society for Engineering Education (ASEE) Annual Conference 2007, Hawaii; http://depts.washington.edu/englib/eld/conf/conf07.php] provides a good visual representation of the engineering design process:

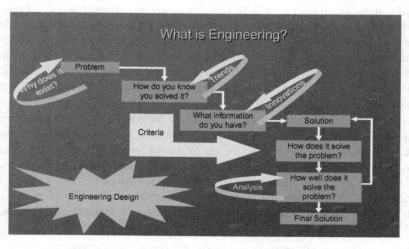

Therefore, it is important to have a basic understanding of the different steps in the design process. There are many books outlining this process; the one by Cross (2000) provides an excellent summary of the different design models. Cross suggests that the design process is a four-stage model: exploration, generation, evaluation, and communication.

According to Cross, French (1985), has a more detailed model adding the following activities: analysis of problem, conceptual design, selection of schemes, embodiment of schemes, and detailing. The whole process begins with the statement of need. French defines the "conceptual design" stage thus:

> *This phase takes the statement of the problem and generates broad solutions to it in the form of schemes. It is the phase that makes the greatest demands on the designer, and where there is the most scope for striking improvements. It is the phase where engineering science, practical knowledge, production methods and commercial aspects need to be brought together, and where the most important decisions are taken.*
> *Cross (2000)*

There are also "prescriptive models" that suggest improvements to the conventional or heuristic models of the design process which emphasize the importance of analysis before the generation of solutions. At the heart of the prescriptive models are the notions of analysis, synthesis, and evaluation. Cross reports that Archer (1984), developed a detailed model: "This includes interactions with the world outside of the design process itself, such as inputs from the client, the designer's training, and experience and other sources of information, etc." Cross (2000, p. 35).

Cross identifies Archer's six types of activity:

- Programming: establish crucial issues; propose a course of action,
- Data collection: collect, classify, and store data,
- Analysis: identify subproblems; prepare performance (or design) specifications; reappraise proposed program and estimate,
- Synthesis: prepare outline design proposals,
- Development: develop prototype design(s); prepare and execute validation studies,
- Communication: prepare manufacturing documentation.

It is obvious from the aforementioned models of the design process that the need for information resources is greatest at the "conceptual design" phase. It is important for librarians to realize this and to proceed accordingly. For instance, in a senior design or capstone project, students will need to use IL skills at the beginning of the semester and will need librarians less as they progress in their projects. In summary, we can make the following generalization on the design process:

- Design problems are by their nature ill-defined or undefined;
- Design projects are subject to evaluation based on a set of goals, constraints, and criteria (often set forth by a client);
- The design process is an iterative process, in which solutions are explored, refined, or abandoned and then other solutions are proposed; and
- The end point or result is the communication (usually in the form of drawings) of the design ready for manufacturing.

Communication and Information-Seeking Habits of Engineers

Understanding the communication patterns and information-seeking behavior of engineers provides us with further insights. "Communication Patterns of Engineers," by Tenopir and King (2004) is one of the definitive works on the subject. In this groundbreaking and much-cited book, the authors cover the subject thoroughly, discussing communication models, engineers' information seeking and use, and factors affecting information seeking and use, and then they give an overview of the major research projects in this field. Many studies over the years have concluded that engineers (in contrast to scientists) traditionally tend to reach for a colleague or for an in-house source of information first, rather than seeking out external sources. Perhaps they now reach for the Internet instead?

Hopefully, the emphasis is changing with the shifts in engineering education and the groundbreaking and sustained work of engineering librarians across the country at their respective campuses. Tenopir and King cite an early study by Holland and Powell (1995) at the University of Michigan to support this assumption. The authors offered an elective on information resources to senior engineering students. They compared the set of students who took the course with the set who did not. It is encouraging to note that those who took the course spent subsequently 50% more time searching for information and reading than those who did not.

The acquisition of information-seeking and critical-thinking skills is crucial for engineers, especially given the notion of the "half-life" of information in engineering. Engineers need to have superior information-seeking skills in order to research the next cutting-edge technology—in fact, they need to be ahead of new developments to keep their own and their company's competitive edge. Developing lifelong-learning skills is at the heart of keeping up with the latest developments. As Goad (2002) said: "innovate or evaporate."

Employers frequently express concerns to faculty that new graduate engineers are well-versed and prepared in technical skills, but lack the professional or "soft" skills. Professor Larry Shuman, a leading engineering educator, defined these skills in his significant coauthored article about the challenging task of assessing the five ABET professional skills, which "include communication, teamwork, and understanding ethics and professionalism, which we label process skills, and engineering within a global and societal context, lifelong learning, and a knowledge of contemporary issues, which we designate as awareness skills" (Shuman, Besterfield-Sacre, & McGourty, 2005). The article was one of the first to review the professional (soft) skills and how they can be taught—or actually, how they can be learned—and how to assess them. The article cites many innovative ways educational institutions are teaching these skills in combination with service learning and global aspects of design projects.

Rodrigues' (2001) excellent article entitled "Industry Expectations of the New Engineer" makes the best case for developing students' skills in IL (if one was needed).

Maybe it should be made into a poster to hang in every engineering librarian's office as a reminder:

> Engineers with solid library research skills will generally produce more thorough reports than those without. The ideal time for the engineer to develop his or her information gathering and management skills is not when entering the corporate world, rather, it is during the engineering education where engineering library resources in staff and collections are virtually always superior to that of the corporate world where library service may be limited or non-existent.

It can be seen that academic librarians have their work cut out for them in terms of developing the information-seeking skills in engineering students while they are on campus. These are skills that they will, hopefully, carry forward to their workplace and so they can continue their pursuit of lifelong learning.

References

Archer, L. B. (1984). Systematic Method for Designers. In N. Cross (Ed.), *Developments in Design Methodology*. Chichester: Wiley.

Charyton, C., & Merrill, J. A. (2009). Assessing general creativity and creative engineering design in first year engineering students. *Journal of Engineering Education, 98*(2), 145–156.

Cross, N. (2000). *Engineering design methods: Strategies for product design* (3rd ed.). Chichester: John Wiley and Sons.

Elliott, M. (2001). The well-rounded IE. *IIE Solutions, 33*(10), 22.

French, M. J. (1985). *Conceptual Design for Engineers*. London: Design Council.

Felder, R. M. (1987). On creating creative engineers. *Engineering Education, 77*(4), 222–227.

Goad, T. W. (2002). *Information literacy and workplace performance*. Westport, CN: Quorum Books.

Holland, M. P., & Powell, C. K. (1995). A longitudinal survey of the information seeking and use habits of some engineers. *College & Research Libraries, 56*, 7–15.

Rodrigues, R. J. (2001). Industry expectations of the new engineer. *Science & Technology Libraries, 19*(3), 178–188.

Shuman, L. J., Besterfield-Sacre, M., & McGourty, J. (2005). The ABET "professional skills"—can they be taught? Can they be assessed? *Journal of Engineering Education, 94*(1), 41–55.

Simonton, D. K. (2000). Creativity: Cognitive, personal, developmental, and social aspects. *American Psychologist, 55*(1), 151–158. Doi:10.1037/0003-066X.55.1.151.

Sternberg, R. J. (Ed.). (1999). *Handbook of creativity*. Cambridge, UK: Cambridge University Press.

Sternberg, R. J., & Dess, N. K. (2001). Creativity for the new millennium. *American Psychologist, 56*(4), 332. Doi:10.1037/0003-066X.56.4.332.

Tenopir, C., & King, D. W. (Eds.). (2004). *Communication patterns of engineers*. Hoboken, NJ: John Wiley and Sons.

Weisberg, R. W. (1986). *Creativity : Genius and other myths*. New York, NY: W. H. Freeman.

6 Designing Information Literacy Assignments

Introduction

This chapter is exclusively concerned with information literacy (IL) in undergraduate engineering since the book is centered around the ABET accreditation program for undergraduate curriculum. We will leave the issue of IL in graduate programs to others. For the sake of brevity, other related issues such as citation styles and plagiarism have not been addressed. A recent article by Eckel (2010) provides an excellent starting point for further study and research on the important topic of plagiarism.

The Academic Librarian's Instructional Environment

Before we consider assignments that embed IL elements, we need to understand the environment within which academic librarians conduct instruction. Frequently—in some cases exclusively—librarians are invited to "guest lecture" by the faculty instructor for one class session. This is usually referred to by librarians as the "one-shot" instructional session. Usually, the class comes to the library's classroom, which has computer workstations, and the librarian delivers a session that targets the assignment or provides a general orientation—often in a 50 minute class session. (It all depends on the duration of the class session, so if the class session is longer, the librarian may have more time.) This is usually categorized as course-integrated instruction. Within the session, the librarian has to demonstrate the library's online resources, including how to download citations, Boolean search techniques, how to evaluate and cite resources, and guidance on developing research topics. Prior to the session, the faculty member and librarian will hopefully have had some interaction on the assignment. Once the session is completed, the librarian may never interact with the faculty for the rest of the semester on that assignment (though some students may contact the librarian for further research assistance). The librarian will not know—unless there is some feedback—if the session was successful in terms of the quality of the student research papers. One way for the librarian to assess the success of the session is for the librarian to analyze the citations from the assignment (see Chapter 7).

However, the current trend and goal of all instruction librarians is to reach out and collaborate with their respective faculty to define learning outcomes, and to design and integrate meaningful and authentic assignments, thereby improving the IL skills of students and their research output. Engineering librarians, bolstered by the ABET Criteria 3i on lifelong learning, have made great inroads in recent years in developing

Lifelong Learning for Engineers and Scientists in the Information Age. DOI: 10.1016/B978-0-12-385214-4.00006-4

partnerships with their faculty to develop assignments suitable for the engineering curriculum (the section that follows highlights some noteworthy collaborations). Many academic libraries have also succeeded in implementing mandatory programs for IL as a degree completion requirement on their campus—this is the ultimate goal of all academic libraries and librarians.

Some excellent research has been conducted on faculty–librarian collaboration. Two highly cited articles are by Leckie and Fullerton (1999) and McGuinness (2006). Leckie points out that in many cases faculty are not aware of what librarians can actually do in terms of library instruction—and it is often the case in engineering. Engineering faculty, in particular, certainly believe in the concept and philosophy of lifelong learning. They would agree that IL is an important skill set to possess for achieving lifelong learning, but they are not sure how to achieve it. One way forward for engineering librarians is to conduct an exercise in curriculum mapping and then to reach out to faculty in a systematic fashion.

Curriculum Mapping

Most courses in the humanities and social sciences require students to conduct research on a topic and then to produce a term paper or some other writing product. Therefore, it is customary for faculty in these subject areas to seek out librarians on a regular basis to teach IL skills to their students to support their assignments. So it is not surprising for most academic libraries (from community colleges to research institutions) to have well-established programs of library instruction in the humanities and social sciences. This is not the case in science and engineering.

The technical nature of many of the courses in science and engineering do not lend themselves naturally to the inclusion of IL, thereby presenting a much greater challenge for librarians in these areas. Engineering and science librarians would be served well by conducting a curriculum mapping exercise to determine the courses that can accommodate IL and then targeting their outreach efforts accordingly.

The Instruction Section of the Association of College and Research Libraries (itself a section of the American Library Association) has published a useful document titled "Analyzing Your Instructional Environment: A Workbook." The section on curriculum mapping provides us a good starting point:

> *A curriculum map provides a holistic view of the integration of information literacy into your institution. To create a curriculum map, list every course offered at your university, then track which courses have a research component, include information literacy, and the content and duration of the information literacy instruction. This process allows librarians and other members of the university community to identify where students are receiving information literacy instruction, the content that is covered, and where gaps exist.*
>
> *http://www.ala.org/ala/mgrps/divs/acrl/aboutacrl/directoryofleadership/ sections/is/projpubs/aie/index.cfm*

This methodology can also be used to analyze the offerings at the college or program level. College catalog entries and syllabi are reviewed for information about the contents of the courses to ascertain if there is a research component. Simultaneously the sequencing of courses is also analyzed. The next step is to note courses that already have an IL component, and then to identify courses that will be suitable for IL (this is the challenging part for engineering and science). After creating the curriculum map, librarians can see at a glance the vertical alignment of courses and can build a program of IL that uses an incremental or building block approach. The analysis allows librarians to create appropriate assignments and to assess what students mastered in one course. Then they can focus on building new skills and knowledge in subsequent course work. This analysis may also allow engineering librarians to be proactive and creative about suggesting a research assignment for a course that may not have had one. A systematic analysis also assists librarians in communicating with department chairs and program coordinators. The most efficient solution is to build and incorporate IL program incrementally in courses that *all* students have to take, rather than randomly teaching sessions if and when asked by faculty. Frequently, campuses may only have two librarians to cover engineering and science, so it pays to be systematic.

In "Designing Better Engineering Education Through Assessment" (Spurlin, Rajala, & Lavelle, 2008), the authors outline the process of curriculum mapping to Outcomes a–k. They describe different ways institutions map their curriculum to the Outcomes. One of the interesting methods uses a numerical system. Part of this chart is reproduced for illustrative purposes:

Curriculum Mapping to ABET Outcomes (Fictitious University)

COURSE/OUTCOME	225	329	499
A mathematics	4	0	4
B experiments	1	1	1
C design*	2	4	4
D teams*	0	0	4
E solve problems	2	1	4
F professional/ethical*	2	1	2
G communicate*	1	4	4
H global*	1	4	4
I Lifelong	**0**	**4**	**2**
J contemporary issues*	2	2	1
K engineering tools	2	4	4

*Reproduced from "Designing Better Engineering Education Through Assessment" (Spurlin et al., 2008) for illustration.
Details about the Outcomes are in Chapter 3 or at www.abet.org.

These numbers refer to the following:

- Major (4): Topics are fully introduced, developed, and reinforced throughout the course. Students have "application knowledge."
- Moderate (2): Topics are introduced and then further developed and reinforced in later lectures. Students have "working knowledge."
- Minor (1): Topics are introduced. Students have "talking knowledge" or awareness.
- (0): Does not relate.

In this fictitious example, faculty would have determined which ABET Outcomes are relevant for these (fictitious) courses. Librarians would primarily be interested in classes that are evaluated by faculty as highly relevant (with a score of 4) for Criteria 3i (lifelong learning) and less interested in classes that score high (for instance) for Criteria A (Mathematics). The scores are determined by the nature of the course. The assumption is that librarians can embed an appropriate information literacy assignment into the curriculum that score high in Criteria 3i. However, even if a course does not explicitly state that it satisfies 3i, but it rates high or moderately high for global and ethical outcomes (for instance), faculty could be approached to develop an assignment that would incorporate IL. Some courses may also have hidden or not immediately apparent ways to use IL: for instance searching for government regulations, information about hazardous substances, or building regulations.

Examples of IL Programs in Engineering (a Literature Review)

Despite the challenges outlined in the previous section, many engineering librarians have developed excellent and well-established programs of IL. The Engineering Libraries Division (ELD) of the American Society for Engineering Education (http://www.asee.org) is one of the best resources for learning about IL programs in academic engineering libraries. Librarians share and exchange information via conference presentations at the Annual Conference: please see the ELD web site, http://depts.washington.edu/englib/eld/, for details. Over the past 5 years, a growing number of presentations have been devoted to the topic of IL. Librarians are also actively publishing about their programs. A few noteworthy examples are as follows.

Arnold, Kackley, and Fortune (2003) describe a successful program for freshmen engineering students at the University of Maryland that evolved from a lecture format to more interactive sessions. Quigley and McKenzie (2003) also report that library instruction for their Technical and Communication classes at UC-Berkeley adapted to the needs of the course by incorporating active learning elements. Nerz and Bullard (2006) report on the program at North Carolina State University, in which engineering librarians developed an integrated instruction program with the Department of Chemical and Biomolecular Engineering to infuse information skills into the curriculum, with assignments and grading. MacAlpine's presentation at the ASEE conference (MacAlpine & Uddin, 2009) with an engineering faculty member describes IL instruction at Trinity University outlining how each element builds on the other. Andrews and Patil (2007) begin with the premise that "[t]he ability to access, evaluate and synthesise high-quality research material is the backbone of critical thinking in academic and professional contexts for Engineers and Industrial Designers." With this in mind, the teaching faculty and librarians developed a program for a first-year unit in "Engineering & Industrial Design Practice" at the University of Western Sydney. The authors conclude that library sessions and assessment tasks were effective (based on feedback and assessment results) in teaching IL skills together with critical-thinking skills. Jay Bhatt and his colleagues at Drexel

University have developed a robust program for undergraduate students. His paper with Roberts (engineering faculty member) (Roberts & Bhatt, 2007), published in the *European Journal of Engineering Education*, describes how during the 2005–2006 academic year, the engineering librarians deployed a new methodology for instruction. They combined an online tutorial for basic library skills and face-to-face consultations between student design teams and engineering librarians. The purpose was to take into account different learning styles together with an active learning component delivered at the students' point of need. It has been shown that delivering instruction at the point of need is more effective than teaching to perceived need and helps students retain information and skills better.

One of the challenges for engineering programs is a crowded curriculum. Faculty are hard-pressed to fit in everything they are required to cover and are reluctant to give up class time for library research—even though they wholeheartedly believe it is an important skill, not just for the program, but also for lifelong learning. To deal with this situation, librarians have developed online tutorials (Aydelott, 2007; Scaramozzino, 2008). Maness (2006) describes the University of Colorado at Boulder's (UCB) implementation of streaming video to support graduate distance programs in engineering and evaluates the use of streaming video applications for IL instruction. Streaming videos offer a viable media for the many distance education programs in engineering that have been established recently. According to the initial evaluation of the UCB project, there is no significant difference between the satisfaction levels and learning outcomes between students who attended in-class sessions and those who watched the streaming video.

Problem-Based Learning and IL

Problem-based learning (PBL) is an important pedagogical tool in engineering education. Some librarians, especially engineering librarians, have been exploring the concepts of PBL and its application to IL.

According to Wikipedia:

> *Problem-based learning (PBL) is a student-centered pedagogy in which students learn about a subject in the context of complex, multifaceted, and realistic problems. Working in groups, students identify what they already know, what they need to know, and how and where to access new information that may lead to resolution of the problem. The role of the instructor is that of facilitator of learning who provides appropriate scaffolding of that process by (for example), asking probing questions, providing appropriate resources, and leading class discussions, as well as designing student assessments.*
>
> *From http://en.wikipedia.org/wiki/Problem-based_learning,*
> *accessed September 26, 2011.*

Characteristics of PBL are:

- Learning is driven by challenging, open-ended, ill-defined, and ill-structured problems.
- Students generally work in collaborative groups.
- Teachers take on the role as "facilitators" of learning.

Hsieh and Knight (2008) report on their experimental project, which incorporated PBL into their IL instruction at the University of the Pacific. They state:

> ... *Many librarians are aware of the potential positive outcomes of employing problem-based learning in bibliographic instruction. This active, participatory teaching style is consistent with the goals of information literacy in several ways. As an inquiry-based form of instruction, PBL closely parallels the Association of College and Research Libraries Information Competency Standards for defining and purposefully resolving an information need. Similarly, both PBL and IL share an overarching goal of instilling skills and abilities for lifelong learning. In addition, the theoretical bases that underlie both IL and PBL recognize and address the influence of learning styles on student acquisition and retention of skills and knowledge.*

The librarians conducted a pilot study comparing lecture-based learning (LBL) with problem-based instruction, including role playing. The results of the reflective survey of the pilot study indicate that using PBL in the early part of education—such as in the freshman year—has advantages over the LBL approach. Students in the PBL section ranked their class higher in interest, participation, and knowledge transfer. In addition, the emphasis on group work and communication also aligns with many of the ABET program outcomes. Hsieh and Knight (2008) also conducted research on the learning styles of engineering students. Most engineering students fall under the Myers–Brigg Type Indicator (MBTI) Introvert, Sensing, Thinking, and Judging (ISTJ). They summarized the groundbreaking research by Felder and Silverman in this area:

> *In their pioneer learning and teaching styles research using multiple personality assessment models, Felder and Silverman identify some important attributes of students with Sensing indicator—they like facts, data, and hands-on learning but may have difficulty with concepts and symbols. They argue that since conventional lecturing uses words and that words are symbols, it puts Sensing students in a disadvantaged position. In addition, they observe a mismatch between engineering students' preferred learning styles and the current instructional styles in engineering education. While most engineering students are visual, sensing, and inductive learners, most engineering education is auditory (lecturing), intuitive (emphasizing concepts), and deductive (principle first, application later, if ever).*

However, PBL is a challenge within the typical 50 minute "one-shot" library instruction session, given that to be done adequately, it requires students to collaborate. Enger et al.'s (Enger, Brenenson, Lenn, & MacMillan, 2002) summary of a preconference workshop at LOEX-of-the West came to the conclusion that class time needs to be at least 75 minutes to facilitate student collaboration.

Critics of PBL suggest that active problem solving early in the learning process is a less-effective strategy than incorporating it later in the process. Early learners find it hard to assimilate large amounts of data and can become overwhelmed (Sweller, 1985). Given these challenges, librarians perhaps should not use PBL in an introductory course, but should use it mid-program, instead. Senior and capstone design projects are by definition using PBL.

Engineering Information Resources

Engineers have to use a vast amount of information and data to function effectively in the workforce. It is outside the scope of this book to go into great depth in terms of individual reference titles in any one of the engineering disciplines; we leave that to the many excellent guides to engineering literature, such as "Using the Engineering Literature," edited by Osif (2011).

All engineers—at the very least—need to be able to find peer-reviewed research articles, trade magazine articles, conference papers, and books. They also need to know the importance of each type of material and when it is important to use a particular format. Articles in trade magazines, not so important in non-sciences and non-technology disciplines, are very important for engineers. At various times, engineers also need to know how to find technical data (mostly found in handbooks), technical standards, patents, and manufacturers' (trade) information. "Gray literature" is hard to define, but includes such formats as technical reports and government reports. Business and company information is important for engineers who want to become entrepreneurs. The Internet has been a boon to engineers, but with information overload comes the importance of critically evaluating information. Web 2.0 resources have added to the complexity of the situation.

Engineers are usually searching for the most recent articles (within the last 5 years is the rule of thumb) on a given topic. However, there are subtle differences and nuances between the various engineering disciplines: civil engineers are more likely to use older material (for instance to study bridge failures), whereas computer engineers generally only use the most up-to-date material. Civil engineers need to know about government publications, whereas they are of less importance to electrical or computer engineers, and so forth. Conference papers and proceedings are extremely important to computer scientists and engineers since their field expands at lightning speed, and the most important developments are often first reported at conferences.

Given the vast array and variety of materials, it is impossible to develop skills in finding and using all the material in *one* library session. A sequenced IL program is the best way to introduce these materials as appropriate for research assignments. The senior thesis or capstone design project is often the place that all these elements are brought together in a PBL environment.

It is a known phenomenon, that without guidance, students tend to search only the Internet for research material for their research papers. Teaching faculty and librarians make good partners—not only in satisfying ABET 3i but also in weaving some of the other Outcomes (communication, teamwork, etc.) into a research assignment.

Choosing Term Paper Topics

Finding interesting and engaging research topics is more of a challenge for the freshmen undergraduate engineering students than for their counterparts in other disciplines given the emphasis on problem-solving and laboratory work in the engineering curriculum. The easy way out for the instructor and librarian is to come up with a set of topics and then ask students to choose from them (Hoover dam, channel tunnel, or more specific topics such as rapid prototyping or SI/versus Imperial units are good examples). Another approach could be short term papers on such topics as the history of invention/inventors or on the theme of "How things work." In a class of 30 students—even working in teams—the situation often occurs that student teams pick the same (easiest) topics. If students are left free to choose their own topic, they frequently choose very broad topics such as civil engineering or aerospace, or they simply struggle with finding a suitable topic, before they can even tackle the task of conducting a literature search.

Faculty and librarians are eager to expose students early in the program to the exciting world of engineering. Finding such a list can be challenging. One method that has worked very well at CSULB (California State University—Long Beach) in the multisection ENGR 101 (course Engineering 100) course is based on the very short articles that are published in the ASEE magazine *PRISM* under the section "First Look," previously called "Briefings," http://www.prism-magazine.org/. *PRISM* is not the only source that publishes short articles on innovative, cutting-edge developments in engineering, but it is used here for illustrative purposes. "First Look" reports on recent, cutting-edge developments worldwide in industry and research institutions. Each of these is packed with information written in nontechnical language, placing it in a societal and global context.

To alleviate some of the issues previously mentioned, in our program we download several articles, categorize them under broad headings (e.g., civil engineering, computer science, energy, alternative energy, transportation, and so on) place them in "folders" in the Learning Management System (LMS) and have student teams pick a topic (once picked it is off-limits to other teams). Almost all of the articles are interdisciplinary (many are in the biomedical and environmental fields) giving students insight into the range of activities with which engineers are involved. This also illustrates to students the importance of their science curriculum and that once they graduate they will be working shoulder to shoulder with professionals in related fields.

The team uses this article as a jumping-off point to research the topic and write a term paper and presentation. They are required to cite at least five peer-reviewed articles in their team term paper. Refreshingly, freshmen engineering students meet the challenge of this assignment well! They have to first understand the article, extract the main points, come up with a set of key words, connect them with Boolean operators and construct search statements, and then search appropriate library databases to find peer-reviewed articles. This process requires a higher level of critical-thinking skills than is usually required for freshmen term paper topics. Since each article is unique, each team has a different set of challenges. An interactive session (often with the use of personal response system, iClickers) is used to discuss the article.

This assignment has been used for three semesters to date and the faculty report that the term papers and presentations are a much higher quality than previously.

(Usually there are 12–15 sections of ENGR 101: 30 students per class and 5–6 faculty members teaching.) In fact, the idea has been floated that a poster session should be organized at the end of the semester with prizes.

Engineering faculty and librarians are finding creative ways to incorporate IL into their engineering curriculum. For instance, Van Treuren (2008) confirms that faculty have no trouble understanding and appreciating the ABET Outcomes covering lifelong learning and societal and global issues, but they have difficulty incorporating these into an already crowded curriculum that needs to cover engineering fundamentals. He incorporated these elements—into a heat transfer class, even—with students making short presentations on contemporary issues related to energy production. Extra credit was given to students who used the research materials available via the library and did not rely solely on the Internet. Riley and Piccinino (2009) describe an impressive program at Smith College incorporating substantial elements of IL into a second-semester first-year mass and energy balance course that uses life-cycle assessment (LCA). The faculty member further reinforced the librarian's role by assigning readings on IL and conducting a class discussion on IL in the broader context of intentional learning and reflective judgment. The course laid the framework for additional IL instruction throughout the engineering core curriculum and in the capstone design clinic.

"Citation Analysis" would be the best assessment tool for traditional term paper assignments (see Chapter 7).

IL in Design Projects

Chapter 5 described the design process in detail, and it was established that the optimum time to insert IL into a student's education is in the conceptual design phase when solutions are being generated and explored and data is being collected. Librarians can enhance the conceptual design phase by pointing out information resources that may spark creativity leading to other avenues and ideas. This could also be called informally the "brainstorming" phase. This section illustrates how IL elements can be incorporated into the conceptual phase of design projects in the engineering curricula from freshman to capstone/senior design projects.

In passing, one should observe that generally librarians are more accustomed to teaching traditional sessions in which they show students how to find citations and research materials for specific topics. These can perhaps be categorized as well-defined or finite problems: the topic is outlined, key words are selected, and databases are searched to find information. This is an important skill in the context of engineering or any discipline. However, design projects are by their nature ill-defined problems with many possible solutions. This presents challenges for librarians, but with collaboration and discussion with engineering faculty, librarians can weave in IL elements using a myriad of information resources to assist in the process. Students must have or develop an understanding of basic engineering principles and then evaluate and utilize a variety of information resources to judge their appropriateness for the task at hand.

Freshmen Design Projects

Elements of the design process are incorporated into different parts of the engineering curriculum in 4-year degree programs and community colleges. As mentioned previously, critical thinking is an essential part of the design process so IL is a natural ally to enhance and support the creative basis of innovation and design. There is some evidence that introducing the design process early in the program is an important tool in student retention. Increasingly, more programs are incorporating a simpler design assignment into the freshmen program, recognizing the fact that many engineering students most likely selected this major because they are creative and like to build or design objects. It is well worth quoting Felder again here: "... not to extinguish the creative spark in our students" (Felder, 1987).

A typical freshmen design project is one using a mousetrap as a source of energy to power a small vehicle or device. Alternatively, faculty frequently purchase commercial design kits and distribute them to students for designing such things as a robotic arm that can pick up (for instance) ping-pong balls. Other simple projects include designing a device to perform a household task or designing an educational toy. A drafting or engineering design course could incorporate a project to design (for instance) a dorm room. A very interesting example—and there are many—is one from Northwestern University in which faculty in the Engineering Design and Communications (EDC) course (a required course for all engineering students) have developed a curriculum based on real-world design problems (Ankenman, Colgate, Jacob, Elliot, & Benjamin, 2006; Hirsch et al., 1998). EDC is designed and taught by faculty from both the engineering school and the university's writing program, and students work in small teams to tackle real-world design problems brought to them by individuals, not-for-profit organizations, entrepreneurs, and industry. Students learn about the design process; about written, spoken, and graphical communication; and about teamwork and collaboration.

MacAlpine (2005) provides an excellent strategy for librarians:

> As is my custom with library instruction, I assumed the role of a student, described a problem that was prominent in my life (excess dog fur all over the house), and did the research to help develop my solution (a vacuum cleaner to use on pets).

So, one of the best ways for a librarian to teach how to incorporate IL into the design process is to come up with their own invention and illustrate how to use library resources—even if the idea is a little far-fetched.

Using MacAlpine's idea, see Appendix for an "invention" by one of the authors (Ramachandran) of this book to illustrate the point. The Appendix outlines a problem statement and a possible solution/invention an inventive (far-fetched or even crazy) idea, but serves a purpose to illustrate the design process and the use of library and information resources. And history shows that some ideas are far-fetched when they are first "invented." The librarian's description is greeted with "rolling eyes," guffaws and smiles—it is alright if it serves a purpose! Library orientations for design projects generate lively discussion as students in the class brainstorm using their critical-thinking skills on the characteristics of materials and components. There are reference materials and handbooks that can be consulted that provide characteristics and specifications for a variety of materials. Below are three examples of handbooks that can provide information on materials with certain characteristics and ideas on how to tether the material to the floor:

* *Handbook of Materials for Product Design*
* *Marks' Standard Handbook for Mechanical Engineers*
* *Grippers in Motion: The Fascination of Automated Handling Tasks.*

A search was conducted on the "invention" in the Proquest database, which was the most appropriate database available at PCC (Pasadena City College); however, any business database, such as ABI Inform, would be the best place to start to determine whether someone has invented something similar to the proposed invention or if there is something similar on the market. In this case we are looking for journal articles, not only in scholarly journals, but also in trade journals and magazines.

Describing your invention is the most challenging part of the exercise. Brainstorming about key words resulted in: umbrella, bubble, and tent, among other descriptions. Proquest revealed that hospitals use something similar as portable operating units for isolating patients; however nothing similar was found for the library market. If we really wanted to pursue our invention, we could approach the manufacturers of the hospital "tents" to find more detailed specifications about their product. Databases in the library field were also searched, since this invention would be for use in libraries. Students were instructed to keep a record of the articles and in particular company names and product details. A search of US patents was also conducted; Google Patents http://www.google.com/patents offers a very user friendly way to search for US Patents. Using a research log would be the best way to keep track of research and can be used for assessment (see Chapter 7). To support such a program to its fullest, more than the typical "one-shot" session would be required (Okudan & Osif, 2005).

To conclude this session, it was gratifying that the PCC engineering faculty member liked this idea so much that he followed up with a "site visit" to the library to analyze the issues with the class.

IL and Capstone Design

The accrediting body for engineering—ABET—requires programs to have a culminating capstone design experience, so almost all 4-year engineering programs in the United States culminate in a capstone or a senior design course. Meyer's introduction in his chapter, entitled "The Capstone Experience at the Baccalaureate, Master's and Doctorate Levels" in "Designing Better Engineering Education through Assessment" (Spurlin et al., 2008), provides an excellent framework for understanding the nature of the capstone experience. He describes the experience in the following way:

> Often called a "capstone design" or "senior design" course, the culminating design experience in an undergraduate engineering curriculum assists undergraduates as they transition from students of theory to practicing engineers. It is in this setting that they begin to truly understand the relationships between different aspects of their chosen engineering disciplines, as well as the necessary balance between scientific theory and design practice. (p. 286)

Meyer quoting Gravander, Neeley, and Luegenbiehl (2004):

> The major design experience needs to introduce students to the messiness of the "real world," a sense of ambiguity, unconstrained variables, and a direct impact on the world's social and economic fabric.

Gesink and Mousavinezhad (2003) give us more clarification: "It would be a project that would require the development of a component, product, or system that has the potential for satisfying a real need."

These courses are rigorous, usually team based, frequently spread over two semesters, and require students to design a product, device, or process and to assess many of the ABET Outcomes a–k. The teams stay together for the two semesters, forming a tight-knit group. It has been suggested that the capstone design course at the end of the 4-year degree program is as close a match as we can find to simulate how a design team functions in industry. The design team experience is the closest facsimile to what engineering students will encounter when they start working as engineers. This experience is even more important for students who were unable to participate in an internship. Given this scenario, there is no better opportunity for librarians to have a final opportunity to teach IL, critical-thinking skills, and create lifelong learners—in short a capstone IL component. To add to the real-life aspect of the capstone experience, engineering faculty around the country have been developing innovative additions to the basic concept. Kasarda, Brand, and Brown (2007) have added a mini-internship and service-learning component. Shaeiwitz and Turton (2003) introduce a lifelong-learning dimension to the experience.

Van Fleet and Hanyak's conference paper (2000) reports on a senior capstone course "Process Engineering" at Bucknell University, which places emphasis on process design. Since 1990, library research has been a significant component of the course, with student teams meeting with the librarian throughout the course. The project involves designing a safe, efficient, and profitable process for the dehydrogenation of ethyl benzene to styrene monomer with a series of written reports on the chosen process design being a major course outcome. As can be imagined, the research process has evolved radically over the course of teaching this course with the advent of technology and in particular the Internet.

Weiner's (1996) article describes a program at MIT (Massachusetts Institute of Technology) that provides us an excellent model for supporting the IL elements of a capstone or senior design project. Weiner describes how a team of engineering librarians at MIT became key components in a capstone mechanical engineering design course. In essence, librarians became members of the team: attended lectures and became "information agents." Using the article as a jumping-off point, some-thing similar was proposed at CSULB as a pilot project.

After some brief discussion with the aerospace faculty member teaching the cap-stone course, it was decided that in addition to the usual orientation that had been done the previous years, the librarian would attend three subsequent laboratory ses-sions and work with each of the teams when they were at the design concept stage. The pilot worked extremely well—even better than was expected. Each team shared with the librarian the synopsis of their project. At the lab sessions, the librarian visited each team and offered her advice and services to them. The information needs of each team were quite varied and depended on their project. Their questions ranged from simple requests (how to place inter-library loan requests) to how to find military specifications, or the market for (example) unmanned aerial vehicles. In addition, the librarian was proactive and guided students to specific areas to research for information as described in the article by Okudan and Osif (2005), even func-tioning at times as an impartial observer interpreting and refereeing issues between team members that had ramifications in terms of IL. Many team members also con-tacted the librarian for assistance outside of these sessions. The librarian witnessed firsthand as students grew immensely in their understanding of key information resources. Yet again it is proof that IL skills develop greatly when they are applied to real-life projects, thereby improving the lifelong-learning skills of engineers. This pilot project continues and will hopefully result in a future journal article.

Entrepreneurship

Capstone, senior, and product development courses lead nicely into the subject of entrepreneurship. Engineers are excellent at innovation but not so adept at selling or marketing their products. Many academic programs have courses in entrepreneur-ship, as well as informal opportunities for developing entrepreneurship skills through clubs, networking groups, and competitions to bridge the skills gap. Ramachandran, Toot and Smith (2002) presented a paper, at the Special Libraries Association Annual Conference, on the development of programs to match the entrepreneurial climate on the California Institute of Technology (Caltech) campus, http://resolver.caltech.edu/CaltechLIB:2002.004 Students, at that time (2000–2001) were starting their own companies and needed to acquire knowledge and skills in business techniques. The engineering and science librarians noticed this trend and started by supporting a newly developed course on eBusiness with extensive web site and library instruc-tion, drop-in workshops for the larger community, and resources for the library. Also at the same time, Caltech and the Art Center were awarded an NSF grant to support an innovative Entrepreneurship Fellowship Program (EFP). The EFP was a perfect intersection between engineering, design, and business (student teams were chosen from Caltech and the Art Center). The librarians worked with the teams over a 9-month period. It was the ultimate experiment in developing entrepreneur-ship and lifelong-learning skills. Librarians observed firsthand how students not only acquired, but also mastered information skills in engineering and business. Many went on to become successful entrepreneurs.

Many librarians have seized the opportunity to work with entrepreneurship courses since it is the perfect place for the coming together of all the skills that engineers need to have to be successful in their chosen profession. Drew (2007) reports on a very successful collaboration between the library and the Collaborative for Entrepreneurship and Innovation at Worcester Polytechnic Institute (WPI). Feeney and Martin (2003) review collaborations between business librarians and science librarians in developing IL courses. They continue to describe their own collaborative program at the University of Arizona, which provides instruction for the Materials Engineering and Design course in which students not only have to design a product, but also have to address cost analysis and market research. At a broader level, many librarians are reaching out to entrepreneurship initiatives on campus, as reported by MacDonald, Pike, and Chung in a special issue of the *Journal of Business and Finance Librarianship* (Pike, Chapman, Brothers, & Hines, 2010). Encouraging and supporting entrepreneurship programs and initiatives is very important for the national and global economy as we struggle with educating the next generation of innovators.

Appendix

New Product Design Course in the Engineering Graphics Program at Pasadena City College

Students had to come up with a novel invention or idea. The class was supported with lectures on the design process and product development.

Problem Statement

A chance comment during a library staff meeting was the spark that ignited the idea. Noise in certain parts of the library was becoming a problem. Many users had understandably commented and complained that they needed more quiet areas in the library in which to study. The PCC Shatford Library is one of the most visually pleasing libraries in the country and winner of the 2008 ACRL award for Excellence in Academic Libraries. However, due to its open design, noises tend to carry and, in fact, are even amplified in certain areas. Structural or architectural alterations are out of the question due to costs involved. So what can the library do to create a quieter environment for the students?

Summary

- Noise in some parts of the library,
- Students need more quiet areas in which to study,

- Libraries have become louder in recent years due to laptops, cell phones, etc., and
- The library has some quiet areas with study carrels, but it is not enough.

Constraints

- The PCC President wants to preserve the look and "line of sight" in the library, so the library is not allowed to add more study carrels.
- There is no budget for architectural or structural changes.

Possible Solution/Invention

Perhaps, we can design something portable—like an "umbrella" or "bubble"—that could be checked out at the circulation desk and put over a study desk to create a quiet environment.

What are the characteristics of the material to be used in the "Umbrella" or "Bubble"?

- Material should be light/portable,
- Should it be opaque or transparent?
- It should allow the person to breathe!
- It should allow in light, but not noise (may be the biggest challenge?),
- The material should be such that it can be collapsed, folded, stacked, and easily inflatable, and
- How will it be tethered to the floor?

References

Andrews, T., & Patil, R. (2007). Information literacy for first-year students: An embedded curriculum approach. *European Journal of Engineering Education, 32*(3), 253–259. 10.1080/03043790701276205.

Ankenman, B., Colgate, J., Jacob, P., Elliot, R., & Benjamin, S. (2006). Leveraging rehabilitation needs into freshman engineering design projects. *113th Annual ASEE Conference and Exposition, 2006, June 18–21, 2006*, Dassault Systemes, HP, Lockheed Martin, IBM, Microsoft, and so on.

Arnold, J., Kackley, R., & Fortune, S. (2003). Hands-on learning for freshman engineering students (computer file). *Issues in Science & Technology Librarianship, 37*.

Aydelott, K. (2007). Using the ACRL information literacy competency standards for science and Engineering/Technology to develop a modular critical-thinking-based information literacy tutorial. *Science & Technology Libraries, 27*(4), 19–42.

Drew, C. (2007). Encouraging use of entrepreneurship information resources: Faculty/Library collaboration. *Paper presented at ASEE Conference.* Retrieved from <http://www.icee. usm.edu/ICEE/conferences/asee2007/papers/1063_encouraging_use_of_entrepreneurship_info.pdf>.

Eckel, E. J. (2010). A reflection on plagiarism, patchwriting, and the engineering master's thesis. *Issues in Science & Technology Librarianship, 62*.

Enger, K. B., Brenenson, S., Lenn, K., & MacMillan, M. (2002). Problem-based learning: Evolving strategies and conversations for library instruction. *Reference Services Review*, *30*(4), 355–358.

Feeney, M., & Martin, J. (2003). The business of science: Cross-disciplinary information literacy in the applied sciences and business (computer file). *Issues in Science & Technology Librarianship, 37*.

Felder, R. M. (1987). On creating creative engineers. *Engineering Education*, *77*(4), 222–227.

Gesink, J., & Mousavinezhad, S. H. (2003). An ECE capstone design experience. *2003 ASEE Annual Conference and Exposition: Staying in Tune with Engineering Education, June 22–25, 2003* (pp. 12351–12364).

Gravander, J. W., Neeley, K. A., & Luegenbiehl, H. C. (2004). Meeting ABET criterion 4— from specific examples to general guidelines. *ASEE 2004 Annual Conference and Exposition, "Engineering Researches New Heights," June 20–23, 2004* (pp. 9675–9683).

Hirsch, P., Shwom, B., Anderson, J., Olson, G., Kelso, D., & Colgate, J. E. (1998). Engineering design and communication: Jump-starting the engineering curriculum. *Proceedings of the 1998 Annual ASEE Conference, June 28–July 1, 1998* (10 pp).

Hsieh, C., & Knight, L. (2008). Problem-based learning for engineering students: An evidence-based comparative study. *The Journal of Academic Librarianship*, *34*(1), 25–30.

Kasarda, M., Brand, B., & Brown, E. (2007). Teaching capstone design in a service-learning setting. *114th Annual ASEE Conference and Exposition, 2007, June 24–27, 2007*, Dassault Systemes, HP, Lockheed Martin, IBM, DuPont, and so on.

Leckie, G. J., & Fullerton, A. (1999). Information literacy in science and engineering undergraduate education: Faculty attitudes and pedagogical practices. *College & Research Libraries*, *60*(1), 9–29.

MacAlpine, B. (2005). Engineering+information literacy=one grand design. *2005 ASEE Annual Conference and Exposition: The Changing Landscape of Engineering and Technology Education in a Global World, June 12–15, 2005* (pp. 5253–5258).

MacAlpine, B., & Uddin, M. (2009). Integrating information literacy across the engineering design curriculum. *2009 ASEE Annual Conference and Exposition, June 14–17, 2009*, Boeing.

Maness, J. M. (2006). An evaluation of library instruction delivered to engineering students using streaming video. *Issues in Science & Technology Librarianship*, *48*, 6.

McGuinness, C. (2006). What faculty think-exploring the barriers to information literacy development in undergraduate education. *The Journal of Academic Librarianship*, *32*(6), 573–582.

Nerz, H., & Bullard, L. (2006). The literate engineer: Infusing information literacy skills throughout an engineering curriculum. *113th Annual ASEE Conference and Exposition, 2006, June 18–21, 2006*, Dassault Systemes, HP, Lockheed Martin, IBM, Microsoft, and so on.

Okudan, G. E., & Osif, B. (2005). Effect of guided research experience on product design performance. *Journal of Engineering Education*, *94*(2), 255–262.

Osif, B. A. (Ed.). (2011). *Using the engineering literature* (2nd ed.). Boca Raton, FL: CRC Press.

Pike, L., Chapman, K., Brothers, P., & Hines, T. (2010). Library outreach to the Alabama black belt: The Alabama Entrepreneurial Research Network. *Journal of Business & Finance Librarianship*, *15*(3), 197–207. 10.1080/08963568.2010.487690.

Quigley, B. D., & McKenzie, J. (2003). Connecting engineering students with the library: A case study in active learning (computer file). *Issues in Science & Technology Librarianship, 37*.

Ramachandran, H., Toot, L., & Smith, C. (2002). *Developing E-Business Information Without a Business School*. California Institute of Technology.

Riley, D., & Piccinino, R. (2009). Integrating information literacy into a first year mass and energy balances course. *2009 ASEE Annual Conference and Exposition, June 14–17, 2009*, Boeing.

Roberts, J. C., & Bhatt, J. (2007). Innovative approaches to information literacy instruction for engineering undergraduates at Drexel University. *European Journal of Engineering Education, 32*(3), 243–251.

Scaramozzino, J. M. (2008). An undergraduate science information literacy tutorial in a web 2.0 world. *Issues in Science & Technology Librarianship, 55*, 3.

Shaeiwitz, J. A., & Turton, R. (2003). Life-long learning experiences and simulating multi-disciplinary teamwork experiences through unusual capstone design projects. *2003 ASEE Annual Conference and Exposition: Staying in Tune with Engineering Education, June 22–25, 2003* (pp. 557–565).

Spurlin, J. E., Rajala, S. A., & Lavelle, J. P. (Eds.), (2008). *Designing better engineering education through assessment: A practical resource for faculty and department chairs on using assessment and ABET criteria to improve student learning* (1st ed.). Sterling, VA: Stylus Publishing.

Sweller, J. (1985). The use of worked examples as a substitute for problem solving in learning algebra. *Cognition and Instruction, 2*(1), 59. 10.1207/s1532690xci0201_3.

Van Fleet, J. A., & Hanyak Jr., M. E. (2000). Engineering design: The information component. *2000 ASEE Annual Conference and Exposition: Engineering Education Beyond the Millennium, June 18–21, 2000* (pp. 2441–2446).

Van Treuren, K. (2008). Addressing contemporary issues, lifelong learning, and the impact of engineering on global and societal issues in the classroom. *2008 ASEE Annual Conference and Exposition, June 22–24, 2008*.

Weiner, S. T. (1996). Librarians as teaching team members in a mechanical engineering senior design course. *Science & Technology Libraries, 16*(1), 3–10.

7 Information Literacy and Assessment

Introduction

Immense strides have been made since the 1980s in library instruction and more recently in the assessment and development of tools to measure its effectiveness. The landmark event was the establishment of the "Information Literacy Competency Standards for Higher Education," approved by the Board of Directors of the Association of College and Research Libraries (ACRL) in 2000 at the Midwinter Meeting of the American Library Association in San Antonio, Texas. It was the first step in defining learning outcomes for information literacy (IL). These standards were also endorsed by the American Association for Higher Education (October 1999) and the Council of Independent Colleges (February 2004). The standards have been discussed in detail in Chapter 4.

The ACRL maintains a web site that tracks research and best practices in this field. The "ACRL Bibliography of Citations Related to the Research Agenda for Library Instruction and Information Literacy," http://www.ala.org/ala/mgrps/divs/acrl/about/sections/is/projpubs/bibcitations.cfm, is a supplement to the "Research Agenda for Library Instruction and Information Literacy," http://www.ala.org/ala/mgrps/divs/acrl/about/sections/is/projpubs/researchagendalibrary.cfm. It provides a list of publications and resources to advance librarians' knowledge of issues related to learners, teaching, organizational context, and assessment. The Bibliography was originally compiled in 2003 and is updated biennially (the last update was in 2010).

Based on the ACRL standards mentioned earlier, the Science and Technology Section of ACRL developed 5 standards and 25 performance indicators for IL in Science & Engineering/Technology to recognize the unique challenges in the science, engineering, and technology disciplines in evaluating, acquiring, and using information. Each performance indicator is accompanied by one or more outcomes for assessing the progress toward IL of students of science and engineering or technology at all levels of higher education (http://www.ala.org/ala/mgrps/divs/acrl/standards/infolitscitech.cfm; also discussed in Chapter 4).

Suffice it to say that the standards provide librarians with a framework and springboard for developing and assessing programs and students. Armed with the ACRL standards, librarians have been actively working on developing learning outcomes and assessment tools. The ACRL standards have been a boon to librarians as a vehicle for communication and discussion of learning outcomes with teaching faculty.

Interestingly enough, for our study, in the same year that the ACRL standards were launched, Engineering Criteria 2000 (EC2000) resulted in the launch of the ABET Program Outcomes or as they are usually referred to as "Criterion 3a–k." Of them, Criteria 3i is the "recognition of the need for, and an ability to engage in lifelong learning" providing engineering librarians an additional vehicle for discussions with faculty on the importance of IL as a vital tool for addressing Criteria 3i. The ABET Program Outcomes are covered in detail in Chapter 3.

Categorizing the different assessment tools is a challenge. Young and Ackerson (1995), reporting on Kirk's 1975 study, summarize his study. It states that research and evaluation of bibliographic instruction (library instruction) methods falls into three main groups (numbered for convenience):

1. The content of the instruction (i.e., whether or not the content of the material presented was learned) as measured by test scores,
2. The process of a bibliographic search (i.e., the steps students go through while researching term papers) as measured by search logs or librarian observation, and
3. The product of the search (i.e., the term paper bibliography) as measured by the quality or quantity of references or the grade received on the paper.

Young and Ackerson (1995) suggest a fourth area:

4. Student self-reporting on attitudes toward the library (as reported in some search log techniques).

Although these studies are dated, they still provide an excellent way of categorizing the evaluation of library instruction and are a starting point for further analysis and study.

As can be seen from the ACRL web site, librarians have been actively researching in this area and have published some excellent books on IL assessment (too many to list here) and we mention only a few as examples or starting points. The ACRL site provides a comprehensive listing. Radcliff's (2007) book is—as its name suggests— a practical guide to IL. It provides detailed descriptions of the major assessment tools, with excellent examples, and then provides the process for analyzing the data. Radcliff provides us a good summary of the whole gamut of assessment tools:

- Informal Assessment Techniques
- Classroom Assessment Techniques (including minute paper, muddiest point, pre-assessment, classroom response systems, defining features matrix, and directed paraphrasing)
- Surveys
- Interviewing
- Focus Groups
- Knowledge tests
- Concept Maps
- Performance Assessments
- Portfolios

Avery's (2003) edited book provides examples for assessing student learning outcomes for IL instruction in a variety of disciplines. Neely's (2006) book "Information Literacy Assessment: Standards-Based Tools and Assignments" provides step-by-step guides to integrating the ACRL standards into assignments.

Jacobson and Xu's (2004) book is an excellent complement to these books and provides tips on motivating students in semester-long IL courses, but also has good tips for all instruction librarians.

Before we consider specific assessment methods, we need to understand the environment within which most academic librarians conduct instruction. (See Chapter 6, section on "One-Shot Library Instruction" for a description of the instructional environment of academic librarians.)

For the remainder of this chapter, we will highlight four assessment tools that are specially suited to the assessment of IL in engineering.

Evaluation of Bibliographies/Citations

In this section we will address the category outlined by Kirk:

- The product of the search (i.e., the term paper bibliography) as measured by the quality or quantity of references or the grade received on the paper.

Ackerson and Young (1994) provide us an excellent survey from the literature (as of 1994) of the different criteria to evaluate the quality and quantity of references in students' bibliographies. Many of the criteria reported by them still apply today, although now librarians have the added challenge of taking into account Internet resources as well as Web 2.0 resources:

- Quantity of sources
- Format
- Currency
- Type of source/variety
- Relevance to topic
- Level of source
- Quality/By recognized authority

Some of these criteria are determined by the requirements of the assignment. For instance, research assignments frequently specify that students can only include n number of Internet sources in their bibliography or can only use peer-reviewed journal articles and they should have been published in the last 5 years. Instructors have had to specify "type of source" since the Internet has become so ubiquitous and left to their own devices, students will write an entire research paper only using Internet resources (supported by evidence and anecdotes).

As an assessment tool, the evaluation of bibliographies (citations) falls within the category of "performance assessments" as defined by Radcliff. It is the systematic assessment of the products of students' performance and all curricula provide us with many opportunities for performance assessments: bibliographies, essays, presentations, research journals, and term papers, among others (figure 11.1 of Radcliff, p. 116).

This helps librarians gauge whether their instruction is on target or whether some adjustments need to be made in subsequent sessions. Engineering librarians in particular have been very successful in collaborating with faculty and this chapter will provide some illustrations.

As collaboration between librarians and faculty has grown, librarians are becoming more comfortable requesting permission from faculty to view term papers and also to assist them in grading the bibliography section of research papers. This must be planned out in advance. The librarian must follow campus guidelines if he or she is officially assisting faculty in grading papers. Faculty can arrange for librarians to review papers *after* the grades have been submitted and the names of the students have been removed. If it is anticipated that this will turn into a research project that will be published, librarians must go through their campus institutional review board (IRB) department.

Some interesting observations can be made about some of the criteria outlined by Ackerson and Young with particular reference to student research assignments in engineering:

Currency:

Generally speaking in engineering we want to include the most up to date (last five years is the rule of thumb) research in an area. This is particularly true in the case of computer science and electrical engineering; however, in civil engineering and mechanical engineering older material can still be valid and appropriate for a research task. Frequently the requirements in terms of date are spelled out in the syllabus. For instance: study of a bridge failure or environmental disaster has to start with the date of the disaster and move forward.

Type of Source/Variety:

The appropriateness of the type of source and variety can vary depending on the discipline and research topic in engineering. Peer-reviewed articles are considered the premier and most important source and are given the highest score in terms of evaluation. Books may be considered the next most important source in terms of providing the "big picture" on a topic. In some engineering disciplines (electrical engineering and computer science in particular) conference proceedings provide the most important information resource next to peer-reviewed articles. Articles in trade magazines are very important in all disciplines in engineering (in contrast to humanities for instance). Government resources are important for the civil engineer but may hardly ever be used by the other engineering disciplines. Standards and specifications are important for all the disciplines. Patents are another important resource especially for design projects. For lower level undergraduates, it is important to stress peer-reviewed journals, books and government resources if appropriate; other types of sources are added as they progress in their academic program ending with the use of all resources as needed in the senior design/capstone project. Graduate students need to understand the breadth and depth of all the resources and its appropriateness for their research.

Quality of Source:

In terms of quality of sources, the well-established criteria can be generally applied to all sources—who, what when. In other fields, students are often instructed to avoid .com or .org sites, however in the case of engineering this is not always possible if one is looking for information about a company, industry or non-profit agency.

Quantity of Sources:

This depends on the type of assignment. Is it a lengthy term paper? Is it a thesis? In those cases we would expect to see a good number of citations. There are some types of assignments in which there may not be many references and that would be acceptable. For instance, a technical briefing on a new or emerging technological development may only have a few references in the literature. Description of a process may have even fewer and those may be in handbooks or other technical material. It may prove to be a challenge to find citations on a case study on an ethical issue in engineering. Design projects bring about their own set of challenges and sometimes the student will have difficulty in finding appropriate sources. The only way the librarian can verify this is by recreating the search themselves or use another assessment tool such as research logs (discussed later in this chapter).

Assessing the Citations

A simple checklist can be used to assess the citations which indicates that the criteria have been met but it does little to measure the quality of the product. A checklist provides better results if a rating scale is added.

To use an example from Radcliff the rating scale for quality of citation style may look like this:

	Superior A	Very Good B	Acceptable C	Borderline D	Unacceptable E
Reference list uses proper APA citation style					

Rubrics are better tools since they "provide a way of measuring the degree to which a particular skill has been mastered, used or incorporated into a bigger process. A rubric detailed descriptions of how separate parts of products or performances will be assessed and assigns a score to each part" (Radcliff, p. 124). A rubric is made up of three components: criteria, indicators of quality and scoring technique. Using an illustration from Radcliff and adapting it to engineering:

Requiring students to provided annotations for their bibliographic entries would help the librarian or instructor in assessing the "Authority" criteria (especially helpful in the case of large classes) as well as help students improve their skills in summarizing or synthesizing information.

Ackerson and Young's collaboration yielded some of the early research in the area. A review of the literature reveals that engineering librarians have been actively developing and refining this assessment method; a few are mentioned here for illustration purposes.

Fei, Sullivan, and Woodall's (2006) project analyzed the bibliographies of first, second, and fourth year chemical engineering students' project reports to identify

	5	4	3	2	1	Score
Currency/ relevancy	Your sources are current (within the last five years) and highly relevant to the topic of nanotechnology	Your sources are current (within the last five years) and mostly relevant to the topic of nano-technology	Most of your sources are current (within the last five years) but a few are questionable relevance to the topic of nanotechnology	Some of your sources are current (within the last five years) and less than half are relevant to the topic of nanotechnology	Your sources are not current (older than the five years) and only a few, if any are relevant to your topic of nanotechnology	
Authority	You identify the author(s) of your sources and they are credible experts in engineering	You identify the author(s) of your sources and most of them are credible experts in engineering	You identify the author(s) of most your sources and some of them are credible experts in engineering	You identify the author(s) of some your sources and a few of them are credible experts in engineering	You identify the author(s) of few, if any your sources and their credibility as experts in engineering is in doubt or unclear	

students' strengths and weaknesses in terms of items cited, the variety of items cited, and the correct use of citation style. The results show that upper level students cited more items (books and journal articles) in total than did lower level students. Not surprisingly, web sites were cited by all three groups, and in the case of first-year students these were the most frequently used sources: the main challenge for all librarians. All students had problems with citation style. The main conclusion is that faculty and librarians have to change the emphasis of sessions from finding to interpreting and citing information.

Gadd, Baldwin, and Norris (2010) reported on an in-depth analysis of the bibliography of 47 final-year projects written by civil engineering students at Loughborough University (England) and found a strong correlation between the grades for the project and the quality of the literature review. They analyze the number, age, and type of references cited, in addition to an investigation into the quality of the bibliographic citations themselves. The authors make recommendations as to how IL teaching relating to the literature review may be improved, mostly in terms of improving the accuracy of citations, impressing upon the students the need to use more current and credible references and to improve the quality of citations. A common problem in engineering is the lack of a common citation style, and the authors make the valid point that academic departments need to develop an in-house style.

Mohler's article (2005) reports the results of citation analysis at Wichita State University of freshman engineering bibliographies to assess the effectiveness of library instruction. Adjustments in instruction were made based on the results of the study. Edzan (2008) reports in a similar study, analyzing the citations of final-year students in the Computer Science and Information Technology department of the University of Malaya.

Denick, Bhatt, and Layton (2010) explored "the assessment of first-year engineering design students' information literacy skills in order to refine existing methods and library instructional strategies." The authors analyze the references cited in terms of citation style, quantity, resource type, and currency of citations in first-year engineering design reports from Drexel University's Introduction to Engineering Design program during the 2008–2009 academic year. From a sample of 234 citations, 38% of references were classified as web sites, 28% of references were journal articles, and 12% of references were books. The results correlate to similar studies. The results of this study were compared to previous assessment efforts and aligned to the ALA/ACRL/STS Task Force on Information Literacy for Science and Technology's Information Literacy Standards for Science and Engineering/Technology.

Citation analysis is an excellent method of assessing the effectiveness of library instruction for all types of instructional sessions from "one-shot instruction" to semester-long projects in engineering. In multisection courses, giving orientations to a few sections and not to others and then comparing the results of the citations may yield some interesting results.

Librarians who decide to do this must make arrangements with the faculty at the beginning of the course so they can view the papers after the grades have been submitted. In an ideal situation, the librarian would assess the citations and it would be part of the final grade.

As librarians, in particular engineering librarians, conduct further research, it will add to our knowledge of student skills and inform and refine our instructional methods.

Research Logs

We have a habit in writing articles published in scientific journals to make the work as finished as possible, to cover all the tracks, to not worry about the blind alleys or to describe how you had the wrong idea first, and so on. So there isn't any place to publish, in a dignified manner, what you actually did in order to get to do the work.
                                     Richard P. Feynman, Nobel Lecture (1966);
                        http://www.gap-system.org/~history/Quotations/Feynman.html

The quotation by Feynman is also true for information research. The librarian cannot see the steps that the student took in conducting research. The librarian cannot be sure that all appropriate resources were used or at the least attempted and what search strategy was used. One of the solutions to remedy this situation is the use of research logs. Research logs would fall in category two according to the framework that Ackerson and Young reporting Kirk's research outlined:

The process of a bibliographic search (i.e., the steps students go through while researching term papers) as measured by search logs or librarian observation.

Librarians have begun to use research logs as a method of evaluation and assessment as can be seen from a search of the Internet; however there appears at present to be a lack of published research in this area.
Tarleton University librarians offer us this information:

Research Logs help researchers formulate searches, modify searches, choose the best search tools, plan and organize research time, document resources, and retrace steps when needed. No matter which search tool is used, keeping a research log that records research activities (sources, search terms, and outcomes) is a good idea. When creating a research log, you should keep track of:

- *dates when you search,*
- *where you search (places and the search tools you use),*
- *search terms and strategies you use,*
- *types of materials you find,*
- *ideas to use during the next research session,*
- *and other notes that will help your research time be more productive.*

This example research log shows the basic categories of information that are usually recorded.
However, this log will not fit all research activities and should be adjusted accordingly.
                                     From http://www.tarleton.edu/departments/library/
                                     library_module/unit3/3log_lm.html

City University of New York offers us another excellent example:

> A "research log" is a record of the process by which your research was undertaken, along with documentation of both positive and negative results; think of the logs as diaries of your research process, listing the sources you consulted (even if they didn't yield anything) and the library resources utilized. To document your research, you should submit a photocopy of the relevant pages of any consulted works (in the case of longer pieces, a copy of the first page of the relevant section will serve), along with the full citation for each source consulted. The log should also include a list of search terms used (in the case of library-based research, a list of used terms as well as the Library of Congress subject designations; for archive-based research, the NUCMC listings as well as information about available in-archive finding guides or related collections; for web-based research, the search terms entered into internet search engines as well as the names and URLs of the engines used). There is no word-limit on logs, but your logs should satisfy the professor that you have a command of basic research methods in history, and that you have sought corroborative evidence for your research even if you don't find much. Your grade on the research logs will be based at least in part on your efforts and your diligence, rather than solely on your results.
>
> From http://www.library.csi.cuny.edu/dept/history/
> lavender/researchlogging.html

To summarize: the main characteristics of a research log are:

- Helps to formulate a search plan,
- Documents the search process,
- Facilitates recordkeeping: important for research spanning several weeks,
- Records the results of searches,
- Records unproductive or unfilled searches,
- Illuminates and reveals gaps in the research plan,
- Records time spent on research,
- Organizes the research process, and
- Ultimately standardizes, organizes, and makes the process efficient.

The research log as an assessment tool is best suited for longer research projects and in particular for design projects in engineering. As mentioned in a previous section, in the case of design classes (lower level and capstone courses), there are often negative and unproductive searches. Students will observe firsthand that research is not linear, leads us in unexpected directions and is frequently iterative. The process mirrors as close as is possible the situation that students will face as working engineers. Hopefully, becoming accustomed to the process as part of the curriculum will translate into good research habits in the workplace.

For the librarian, research logs offer a rich area for assessment. The librarian has an opportunity to observe the critical-thinking skills (or lack thereof) that students employ in the research process. What key words and subject terms did they use? What resources did they use or try? Which resources did they prefer? How much time did they spend on tasks? As mentioned, on the CUNY web site, students' grades should be based on their diligence not on the results of their search. Librarians should look at the contents of the research logs in a holistic way—did the student or student team understand the process of information research? A well-documented research log will reveal this.

Research appointments with the librarian or librarians could be required as part of the assignment to reinforce and guide the research process of students or student teams. However, this depends largely on logistics in terms of librarian workload. Of course, librarians are available for individual consultation whenever students require it. Rubrics need to be developed to evaluate research logs.

CSULB Experience

One of the authors (Ramachandran) has been collaborating closely with a mechanical engineering faculty (Professor Panadda Marayong) at CSULB to incorporate IL into freshmen design projects. Many different projects have been assigned as part of the curriculum for MAE101B since 2008. A good example (for instance) was to design and build a household appliance that is powered by a mousetrap. Since spring 2010, a research log has been used by student teams to record their research. Students have to produce a short paper to support their project and presentation; and they are required to cite at least five references of which only one can be an Internet site. The research logs are part of the final grade. An initial analysis of the data has already yielded some interesting results and suggests a need for some adjustments to the librarian's lecture. For instance, searching for "banana peeler" (the invention) in an information resource will (most likely) result in no hits but if the banana peeler uses the principles of a "pulley" then searching for information on pulleys will give students some useful information and results. Students are also asked to rate the resource producing some interesting results. The ability to use critical-thinking skills to extrapolate and describe the invention in terms of engineering principles is vital to conducting appropriate research and is essential in product development, which is the heart of the engineering field. The faculty member and librarian are hoping to conduct an analysis of student citations and an evaluation of the use of research logs and publish their findings.

Worksheets

Worksheets are a simple and useful tool for assessment of student skills and the effectiveness of library instruction. Worksheets are usually used by K-12 teachers, but they can be adapted for use in college library instruction, especially for freshmen and lower division courses. A worksheet is a one or even half-page sheet that gives students step-by-step instructions on how to do research (e.g., pick a topic, create a list of key words, construct a Boolean statement, search the library catalog, and search databases). It is frequently utilized during or immediately after a lecture and demonstration to reinforce basic principles. It is a known fact that students have difficulty remembering and reproducing the steps in the research process and the use of a worksheet in the hands-on portion of the library session helps the student to apply what they have just observed. Time permitting, it is easy to employ the use of worksheets for the "one-shot library instruction" sessions" and allows students to start their research right away. Worksheets, simple as they are, offer a lot of advantages in terms of assessment for the librarian and faculty. Even an informal observation of the worksheets as students complete them during the hands-on portion of the instructional session will give librarians some immediate feedback and will facilitate one-on-one guidance and instruction and possibly some further demonstrations to the class based on observance of common mistakes. Collaborating with the teaching faculty, it can be used as an in-class graded assignment or a take-home assignment contributing to the final grade. The worksheets can be easily adapted for different classes at very short notice.

Griffin and Ramachandran (2010) describe the successful use of worksheets (adapted from those used at Pasadena City College) for an in-class assignment in an IL program for preservice science education teachers.

An example of a worksheet from the University of Nevada, Reno (although it is called Research Log) is given in the following web page: http://www.knowledgecenter.unr.edu/instruction/help/worksheet.html.

Portfolios

According to the author of "A Practical Guide to Information Literacy Assessment for Academic Libraries" (Radcliff, 2007), portfolios have the following characteristics:

- Provide an in-depth assessment of how IL competencies are applied,
- Assess skills over time,
- Allow for individual differences in expression,
- Require long-term access to students,
- Require collaboration between student and instructor,
- Require high effort on the part of the student (creation) and instructor (evaluation), and
- May involve complex infrastructure if electronic portfolios are used.

The chapter by Radcliff offers us a good summary of the characteristics of portfolios and guidance on how to implement them. They have been used for some time in the K-12 context, but are increasingly being used in higher education at the program or institution level. They are a better representation of a student's accomplishment than what is reflected in grades. Many institutions allow students to access their electronic portfolios for some time after graduation so that they can use these when they apply for jobs. Portfolios are—in essence—a collection of a student's work (or artifacts as they are sometimes referred to) around a theme: in our case it would be the theme of IL. One common element is the inclusion of one reflective piece. Although portfolios have certain common characteristics, there is not a standard model, and it can be for a single class or it can be a record of the student's learning over a period of time. The elements that go into a portfolio are determined by the instructor, together with the librarian(s)— the case of IL—and the requirements of a program or institution.

As Radcliff outlines, librarians become involved with portfolios when their faculty collaborator is using this form of assessment in their course, when the program for which they are a liaison requires it, when the institution has made IL assessment a component of student assessment (oh happy day!), or when the librarian is teaching an IL or library skills course. If portfolios are used for assessment, one will have to decide the context and the exact components.

The number of products in a portfolio is generally between four and eight. The following are types of assignments that could possibly go into an engineering portfolio:

- Annotated bibliography on a topic,
- A research paper (10–15 pages), which would include a reflective piece on the literature review process,
- A research log tied to a design project,
- A technical poster presentation (in power point) on an engineering topic,
- "Technical briefing" on a new/cutting-edge technology,
- Results of a patent search,
- Information on three companies of interest in the industry in which the student wants to work, and
- A reflective journal or logbook. Critique of articles found in the literature on current issues in engineering and science (number of articles to be determined by the instructor).

All of these activities require the student to use information-seeking skills in a variety of different ways. For each of the activities there should be clear guidelines on what is required: the objectives for each element and rubrics. Students will be given opportunities at different points in their curriculum to work on each of these elements and to finish work on them in a final synthesis-type course.

For the undergraduate engineering program, the use of portfolios offers interesting possibilities. As engineering librarians working collaboratively with their faculty map out a sequence of sessions in which to embed IL assignments into the undergraduate program, they could suggest to the program directors that portfolios be used to assess ABET Criteria 3i. The Portfolios could be used for accreditation purposes to demonstrate the range of activities that students have engaged in to satisfy Criteria 3i. The final assessment could be part of a senior design/capstone course or in a professional seminar course.

There are many examples of the use of portfolios in the humanities and social sciences curriculum, but more recently it is being adopted in the engineering curriculum. The Picker Engineering Program at Smith College—based on the ACRL standards and integrating the ABET standards and assessments—uses electronic portfolios in addition to other assessment tools (Riley, Piccinino, Moriarty, & Jones, 2009). A research report titled "Direct Assessment of Information Literacy at NJIT: A Portfolio Assessment Model," http://library.njit.edu/docs/ILassessmentresearchreport.pdf, provides a good summary of the topic.

Finally, one of the best articles to date is "E-Portfolios and Blogs: Online Tools for Giving Young Engineers a Voice" (Carroll, Calvo, & Markauskaite, 2006) bringing us full circle to the use of e-portfolio tools in the pursuit of lifelong learning:

> For the engineer, information literacy means being aware of matters of concern to the engineering profession that appear daily in various forms of media, such as professional journals, magazines, television and the Internet. It includes capabilities to access, locate, manage, integrate and evaluate critically relevant information resources as well as to use ethically and create own information. In terms of communication capabilities, the engineer must be able to discuss these issues, and present their views in a logical and yet forthright manner with their clients, colleagues, and members of the community.

Students in the ENGG1803 Professional Engineering course maintained a reflective logbook as a part of their assessment of an "active intellectual monitoring and evaluation of one's own formal learning and professional practice activities, to examine them for new understandings, and to add to the individual's accumulated knowledge and experience." Students achieved this goal by searching and finding an article related to a specific topic in engineering, thereby improving students' IL. This assignment enabled students to gain a better understanding and appreciation of the various forms of relevant information, including technical books and reports, research articles, and magazines. A written, concise assignment reinforced the significance of the article and its importance to the engineering profession.

References

Ackerson, L. G., & Young, V. E. (1994). Evaluating the impact of library instruction methods on the quality of student research. *Research Strategies, 12*.

Avery, E. F. (Ed.). (2003). *Assessing student learning outcomes for information literacy instruction in academic institutions.* Chicago, IL: Association of College and Research Libraries.

Carroll, N. L., Calvo, R. A., & Markauskaite, L. (2006). E-portfolios and blogs: Online tools for giving young engineers a voice. *Information Technology Based Higher Education and Training, 2006. ITHET '06. 7th International Conference* (pp. 1–8).

Denick, D., Bhatt, J., & Layton, B. (2010). Citation analysis of engineering design reports for information literacy assessment. *2010 ASEE Annual Conference and Exposition, June 20–23, 2010.*

Edzan, N. N. (2008). Analysing the references of final year project reports. *Journal of Educational Media & Library Sciences, 46*(2), 211–231.

Fei, Y., Sullivan, J., & Woodall, L. (2006). What can students' bibliographies tell us?—evidence based information skills teaching. *Evidence Based Library and Information Practice, 1*(2), 12–22.

Gadd, E., Baldwin, A., & Norris, M. (2010). The citation behaviour of civil engineering students. *Journal of Information Literacy, 4*(2), 37–49.

Griffin, K. L., & Ramachandran, H. (2010). Science education and information literacy: A grass-roots effort to support science literacy in schools. *Science and Technology Libraries, 29*(4), 325–349.

Jacobson, T., & Xu, L. (2004). *Motivating students in information literacy classes*. New York, NY: Neal-Schuman Publishers.

Mohler, B. A. (2005). Citation analysis as an assessment tool. *Science and Technology Libraries, 25*(4), 57–64.

Neely, T. Y. (2006). *Information literacy assessment: Standards-based tools and assignments*. Chicago, IL: American Library Association.

Radcliff, C. J. (2007). A practical guide to information literacy assessment for academic librarians: *Westport, CN*. Libraries Unlimited.

Riley, D., Piccinino, R., Moriarty, M., & Jones, L. (2009). Assessing information literacy in engineering: Integrating a college-wide program with ABET-driven assessment. *2009 ASEE Annual Conference and Exposition, June 14–17, 2009*, Boeing.

Young, V. E., & Ackerson, L. G. (1995). Evaluation of student research paper bibliographies: Refining evaluation criteria. *Research Strategies, 13*.

8 Internships

The generic term "internships" actually includes several other categories, and all—in this Internet age—are turning out to be of historic interest.

Cooperative Education

Cooperative education is a term used to describe educational patterns whereby a student alternates spending one semester at a college or university with one semester in the workplace as an "employee" or "student employee." Another pattern adopted is for a student to be less than a full-time student and to work part-time with an employer throughout the academic year.

Both patterns have three characteristics:

- A level of interconnection between the academic program and the work in industry,
- Some type of report or essay on the work experience to be written by the student and graded by a faculty advisor and/or a coordinator of the internship/cooperative education program on campus and/or the work site supervisor of the student, and
- An increase in the length of time needed for the student to complete the degree program at the educational institution.

In the United Kingdom, the term "sandwich course" is used in place of the terms cooperative education or co-op education.

Internships

Internships represent work experience in the chosen academic discipline of the student part-time during the academic year and/or full time during the summer. There is not a great deal of formal connection between the students' academic department (or campus) and the place of employment.

Service learning generally occurs at the course level and the work (read "service") done by the student is the learning experience usually to achieve a learning outcome from the course and—in the rare case—from the program.

Lifelong Learning for Engineers and Scientists in the Information Age. DOI: 10.1016/B978-0-12-385214-4.00008-8

History of Cooperative Education

The history of cooperative education begins with Professor—later President—Herman Schneider of the University of Cincinnati. The institution gained renown for the development of cooperative education (Smollins, 1999). Later, Northeastern University developed a strong co-op–based curriculum that—at one time—became the core of its educational mission. The formation of the Cooperative Education Association, now called the Cooperative Education and Internship Association was an important event. A Wikipedia article on cooperative education (Wikipedia, 2011) gives the rich history in detail. An article by Patricia Cross (1973) also summarizes various issues related to cooperative education.

Core philosophies

The intellectual underpinnings of cooperative education have been well summarized by Kolb (1984). The three main philosophers were John Dewey, Kurt Lewin, and Jean Piaget. John Dewey, an educational and social reformer with his theory of instrumentalism, thought of truth as a changing entity based on the problem to be solved at any time, much in the manner of Buddhism. He sought to bring about a unity between the processes of education and actual experience, thus laying the intellectual groundwork for experiential learning. Kurt Lewin, the father of social psychology, contributed through his research on action research and laboratory training methods. Finally, Jean Piaget, the rationalist philosopher from France, described how intelligence shapes experience. Other notables who have added depth to the theories are Carl Jung, Erik Erikson, and humanist Carl Rogers.

Bridging the Gap Between Academia and Practice

The standard practice in cooperative education is to assign the student intern an essay-writing exercise in addition to the learning assessments. Since co-op education is primarily based on supplementing classroom and lab learning in academia with practical learning through an internship, any instruction related to the enhancement of the co-op education needs to not only add to the academic knowledge but also assist the students in the experiential learning process. Any instruction based on the bedrock philosophies of co-op education should be a safe method of enhancing the co-op education process.

A short 15 hour instruction program for engineering co-op students at California State University, Long Beach, consisted of (Naimpally, 1996):

1. Describing engineering careers available to students. A description of engineering careers is available from books by Kemper (1982), as well as recently published books by Landis (2000). Similar books are available for other disciplines in science, math, and engineering. In this technological age, web sites of professional associations for various

disciplines—biology, chemistry, physics, computer science, and electrical engineering, to name a few—describe the current careers available to students. The US Department of Labor web site describes the current demand for various professionals, and state government web sites in the United States give data relating to the demand for different categories of professionals by region, county, and city. Most industrialized countries give similar information on their web sites. Requesting students give a short 15 minute presentation on their discipline would be useful for freshmen students in a co-op program.

2. During the second part, students gave 15 minute summary presentations encompassing:
   - The task aspects of their internship experiences. This part did not include any teamwork or other social issues.
   - A description of academic learning in their courses at the campus on topics related to their internship experience. If the student is a beginning student, then he/she could request to meet with his/her department undergraduate advisor or any faculty mentor from his/her home department in order to obtain information on topics covered (related to the internship experience).
   - A summary of how the student was able to integrate the above two aspects of their education: the internship experience and the course-based learning.

The second part is useful in assisting the student with integrating information learned and with transitioning from academia to the workplace. It also assists the student with avoiding the formation of mental silos with respect to academic course work and experiential learning. Building a sound bridge is the key to making the student an effective lifelong learner who can integrate and build upon the continuum of seemingly disparate experiences.

3. Instruction in specialized topics related to social issues faced by students in their first experience at their professional workplace:
   - Teamwork issues: Teamwork is essential to the modern workplace. It cannot be denied that our modern system of education—with its emphasis on grades—has encouraged an individualized competitive spirit in students right from the time they receive their first grade in elementary school. Instruction was then given based on a chapter on the dimensions of the group process from the book by Fisher and Ellis (1990).
   - Issues related to protocols at meetings: the young intern or co-op student is quickly faced with meetings at work, whether it be individual ones with the supervisor or coworkers, or group meetings with the department or division chaired by a first- or second-level supervisor. Add to it informal or working-group meetings, and the co-op student could not be blamed for being confused as to the protocol at these meetings. Munter (1992) has described different speaking situations and the protocol to be followed in them. Although the book is written for managers, it would be useful for interns, co-op students, or young employees.

Most academic programs for scientists, mathematicians, and engineers include courses on speech or public speaking with a substantial component that includes delivering speeches in the classroom. Therefore, unless the undergraduate program at an educational institution does not include a speech course, it is not necessary to include instruction in public speaking at a short instructional program.

Bureaucratic Aspects of the Internship Program Office at the Educational Institution

Prior to the self-placement methods promoted through the use of the Internet—whereby students can submit resumes and set up interviews, themselves—the internship or co-op program office at a college or university was generally staffed with professional counselors and clerical support staff. Their job functions included making sales pitches to local and national industries, businesses, and government organizations for promoting co-op arrangements between the external firm and the campus co-op office; counseling students; publicizing the co-op program; and assisting employers by setting up campus interviews followed by on-site interviews. All of these functions—with the exception of the counseling of students—have now become automated. The counseling function can be carried out by counselors from the campus placement office who take on potential co-op students to their workload, in addition to counseling new or upcoming grads about their academic programs. In the case of the larger offices being used for co-op functions, program evaluation of the office functions can be carried out using the principles enunciated by Posavic and Casey (1997).

References

Fisher, B. A., & Ellis, D. G. (1990). *Small group decision making.* New York, NY: McGraw Hill.

Kemper, J. D. (1982). *Engineers and their profession.* New York, NY: Holt, Rinehart and Winston.

Kolb, D. (1984). *Experiential learning.* Englewood Cliffs, NJ: Prentice Hall.

Landis, R. (2000). *Studying engineering.* Los Angeles, CA: Discovery Press.

Munter, M. (1992). *Guide to managerial communication* (3rd ed.). Englewood Cliffs, NJ: Prentice Hall.

Naimpally, A. (1996). A Cooperative Education Program at CSU, Long Beach. Presented at the *California Cooperative Education Conference, San Diego, CA, April 1996.*

Patricia Cross, K. (1973). *The Integration of learning and earning: cooperative education and non-traditional study.* Washington, DC: ERIC Clearinghouse.

Posavic, E. J., & Casey, R. G. (1997). *Program evaluation methods and case studies.* Englewood Cliffs, NJ: Prentice Hall.

Smollins, J. P. (May 1999). The making of history: Ninety years of Northeastern Coop. *Northeastern University Magazine, 24*(5).

Wikipedia. (2011). *Article on Cooperative Education.* Retrieved from <http://en.wikipedia.org/wiki/Cooperative_education>.

9 Learning Contract

Internship Learning Contract

The internship learning contract is a formal contract between the student intern, the (employment) site supervisor, and the faculty internship sponsor/coordinator/advisor.

For examples of learning contracts, see Appendices (Guidelines for Field Experience/Teaching Practicum/Internship Learning Contract, 2011; Internship Learning Contract; Internship Learning Contract, 2011; Internship Learning Contract Tips, 2010–2011; Learning Contract and Internship Proposal, 2011). The basic information required in a learning contract, related to work assignments, is

- Learning objective(s) numbered 1–3. Although some institutions require four objectives, that might—at times—exceed the capacity of the intern due to the newness of the field of experience.
- Each learning objective is further explained: Details of what should be learned through the internship are included such as developing skills, expanding knowledge, testing theories, explaining career interests, and discovering strengths and weaknesses (Internship Learning Contract, 2011).
- How does a student meet the learning objectives? The student can work toward achieving the objectives through a combination of some or all of the following:
  a. Participating in training programs offered by employers or manufacturers of instruments used by employers,
  b. Observing a senior staff member,
  c. Working on a series of progressively more difficult and more complicated assignment(s)—in which each assignment is evaluated prior to commencement of the next assignment, and
  d. Performing a regular "employee-like" assignment—if the student is already proficient in technical skills.
- Learning assignment report(s): A short- to medium-sized report on each learning objective would be desirable, with the written report being orally presented to a "committee" consisting of the site supervisor and faculty sponsor.
- Evaluation: The evaluation of learning objectives would ideally be done by the site supervisor, the faculty sponsor, or both.

The faculty advisor should meet with the intern and evaluate the internship opportunity to ensure the internship learning objectives meet the academic goals of the student intern. Information about the student, for example: name, address, contact information, and faculty advisor's contact details should be included in the learning contract.

Lifelong Learning for Engineers and Scientists in the Information Age. DOI: 10.1016/B978-0-12-385214-4.00009-X

Details about the employer should indicate name of company, supervisor's name and contact information, weekly hours, and amount of credit given to the student.

Since employers generally contact colleges and universities through the career office/center on the college campus, a memorandum of understanding or similar agreement should be signed by the employer and the career center, preferably based on a model set by a national organization like the National Association of Colleges and Employers (Internship Program Brochure, 2011; Principles of Professional Practice for Career Service and Employment Professionals, 2011). Such an agreement would serve to lessen or prevent any abuses involving the use of student interns for routine, noncareer-related tasks during the internship.

Other Issues Related to the Internship Learning Contract

1. Obtaining information about what is needed for developing a learning contract can be done by a "competency method." The step-by-step process involves
   a. The student understanding the components of a successful practitioner in a given science or engineering field, including skills, knowledge, level of creativity, and level of innovation.
   b. The faculty advisor discussing strengths and weaknesses with the student so he or she can understand the student's present level of competency. The "gap" is identified by the advisor.
   c. The faculty advisor discussing the gap with the work supervisor and then jointly determining the extent to which the given internship presents an opportunity to close the gap.
   d. The faculty advisor discussing the conclusions with the student and then jointly coming up with a learning contract that fits the bill and is acceptable to the student. Acceptance by the student is important since a significant percentage of learning in an internship is self-directed, even when it appears to be directed by others.
2. A learning contract can be used to personalize learning experiences. In developing the learning contract, it is important to have a sense of the best approach to be used (Distance Learning Information Packet, 2011) (Table 9.1).

Table 9.1 Your Learning Style Preferences

	Self-Directed Learner	Other-Directed Learner
Learner Dependent	Standard contract with suggested structure used as basic guide	Standard contract using instructor suggestions
Learner Independent	Create own contract in terms of content and procedure	Develop own version of contract using instructor suggestions

Appendix 1

Internship Learning Contract from Gallatin School of Individualized Study, New York University

 | GALLATIN SCHOOL *of* INDIVIDUALIZED STUDY | NEW YORK UNIVERSITY

INTERNSHIP LEARNING CONTRACT

The function of this contract is to establish an agreement among the student, the internship site supervisor, and the adviser on the purposes and logistics of the student's internship. Together, the three parties should complete all sections of this form, attaching additional pages to answer the questions on Parts II and III. This form must be signed by the student, the supervisor and the faculty adviser. Each should keep a copy, and the original should be submitted to Faith Stangler in the Gallatin office.

PART I: LOGISTICS

STUDENT INFORMATION (PLEASE PROVIDE LOCAL CONTACT INFORMATION)

NAME   SEMESTER & YEAR OF REGISTRATIO

UNIV ID N   PHONE

ADDRESS   NYU E-MAIL ADDRE   @nyu.edu

                                                    ADVISER

INTERNSHIP DETAILS

NAME OF ORGANIZATION

ADDRESS

        Supervisor's Information   Intern's Information

NAME   TITLE

TITLE   TOTAL WEEKLY HOURS IN INTERNSHIP

PHONE   NUMBER OF INTERNSHIP CREDITS

E-MAIL   DATES OF NYU HOLIDAYS OR VACATIONS

PART II: THE INTERNSHIP (ATTACH ANSWERS SEPARATELY)

A. Job Description: Describe in as much detail as possible your role and responsibilities in the internship. Identify your duties, any projects that you will undertake, teams you will work with, products or services you will provide, clients/patrons you will serve, etc.

B. Supervision: Describe the supervision you will be provided at the internship site. What instruction, assistance, guidance and consultation will you receive? From whom? Will you have regularly scheduled supervisory sessions?

C. Evaluation: By whom will your work performance be evaluated?

PART III: LEARNING OBJECTIVES AND ACTIVITIES (ATTACH ANSWERS SEPARATELY)

A. Lea cribe in as much detail as possible what you hope to learn through the internship. Be specific: are you talking about developing
skills, ledge, testing theories, exploring career interests, discovering your strengths and weaknesses, or some other goals? Are these
object ea of concentration? If yes, how?

B. Learning activities and strategies: Describe in detail the specific processes by which you will achieve these goals. On-the-job: How will your internship activities enable you to meet your learning objectives? Include projects, research, report writing, conversations, etc., which you will do while working, relating them to what you intend to learn. Off-the-job: How will you supplement the work experience with reading, research and consultation?

C. Evaluation: How will you determine whether you have met your learning goals? By what criteria will your supervisor assess your performance at the internship site? By what criteria will your adviser assess your performance in the internship?

(continued on back)

GALLATIN SCHOOL *of* INDIVIDUALIZED STUDY   TELEPHONE: (212) 998-7370
715 BROADWAY . NEW YORK, NY 10003   FAX: (212) 995-4150

PART IV: AGREEMENT

The undersigned agree to the terms of this learning contract.

_____

Student's Signature   **Date**

_____

Supervisor's Signature   *Date*

_____   _____

Adviser's Signature   *Date*   *Adviser's E-mail Address (or best way to contact)*

Appendix 2

Internship Learning Contract Tips, University of Minnesota

Internship Learning Contract Tips

Career & Internship Services
CCE • CDes • CFANS

UNIVERSITY OF MINNESOTA

198 McNeal Hall (St. Paul) • 411 ST&SS (Minneapolis) • 612-624-2710 • www.careerhelp.umn.edu • careerhelp@umn.edu

Learning Contract Defined

The learning contract serves as an outline for you and your supervisor of what you intend to learn and accomplish while you are at your internship. It includes your planned activities and projects, your shared intentions and expectations, and the roles you and your supervisor will have working together.

Suggested Guidelines

Begin with a perspective that you are making a contract with yourself. You are identifying what knowledge, behavior, competencies, attitudes, and values YOU wish to develop. These learning objectives are YOUR plan (not your site supervisor's, nor your internship advisor, nor your parent's) that outlines how you will attempt to reach your goals. The following is a step-by-step guide to successfully starting and completing your learning contract. *Adapted from Messiah College Internship Center Materials

Step One: Identify learning objectives most relevant to you.
- Reflect upon your prior educational and life experiences.
- Consider your future aspirations. What will move you from where you are currently to where you desire to be?

Step Two: Brainstorm responses to a few key questions.
- "What do I most want to explore, understand or learn during my internship?"
- "How would I like to change or be different by the end of my internship?"
- "What will make me more marketable to an employer or graduate school?"

Step Three: Decide which goals are most related to this internship.
- You have a role to fulfill at the organization you are interning. You want to be sure to meet their needs and expectations.
- Be clear with your supervisor about what you are looking for from this experience as well. It should be a give and take. If a stated aspiration of yours is not necessarily going to be possible, talk with your supervisor about experiences related to your desired goal that might be possible.

Step Four: Prioritize the list of goals.
- Which ones are most important to you?
- Do the objectives support academic, professional and personal concerns?

Step Five: Prepare the first draft of your learning objectives.

- Utilize the above brainstorming to write a formal list of your learning objective that address your intended goals and outcomes.

Step Six: Identify activities that will help you reach your objectives.

- What work activities and assignments will help you reach your objectives?
- What resources outside of the work site may help you reach your objectives?
- Consult with your internship advisor, internship site supervisor, co-workers, departmental faculty, peers, and the "Learning Objectives Tip Sheet" for ideas.
- Attempt to quantify where possible (e.g. Read 2–3 journal articles).

Sample Learning Objectives

Learning Goal 1: To observe and better understand youth development and the development, implementation, and evaluation of leadership, community service, and recreational activities for youth.

To be met through the following tasks:
1. Observe and interview staff members who directly are a part of youth development,
2. Observe, firsthand and through documents, the marketing of youth programs and activities,
3. Observe and participate in building relationships with community partners that provide youth programs,
4. Find/read three current articles on youth development in academic journals.

Learning Goal 2: To apply and further develop skill in the teaching of ecology and conservation of aquatic habitats pursued through class work.

To be met through the following tasks:
1. Receive training in developing lesson plans, scheduling programs, and evaluation,
2. Give programs during the summer to school age youth, preschool children, seniors, and people with disabilities and diverse cultural backgrounds,
3. Keep program records and enter them into a computer database,
4. Spend 2–3 days assisting a Fisheries Biologist in the field.

Learning Goal 3: To possess a clear understanding of online merchandising, ultimately assisting in the development and execution of online strategies across various dot com merchandising departments.

To be met through the following tasks:
- Conduct research and analysis of the company's online retailing and other successful online retailers,
- Work with merchants, marketing teams, and other critical team members to improve departmental planning and reporting around online marketing initiatives,
- Conduct analysis of cross-departmental product lifecycles and seasonal sales,
- Keep a running journal recording the information obtained as well as skills or tasks yet understood fully.

Appendix 3

Learning Contract and Internship Proposal, State Univeristy of New York, Plattsburgh

LEARNING CONTRACT & INTERNSHIP PROPOSAL
SUNY Plattsburgh

INSTRUCTIONS
1. Prepare a preliminary draft of this learning contract and internship proposal. (See Sections II, III, and IV on reverse.)
2. Discuss your duties as an intern with your site supervisor.
3. Present your transcript (available on Banner Web), your draft proposal, and this Learning Contract to your faculty sponsor for discussion.
4. After the faculty sponsor approves your draft, prepare a final version and obtain signatures. (Section V on reverse.)
5. Submit this form to the dean of your sponsor's faculty. If approved by the dean, copies of this form will be distributed to: student, faculty sponsor, and department chairperson.

GENERAL INFORMATION

Student's Name:_____ Student's ID:_____

Local Address:_____ Phone No.:_____

Email Address:_____

Address During Placement:_____

_____ Phone No.:_____

Major(s):_____ Class Level:_____

Internship Title:_____ Course Number:_____

Date internship begins:_____ Date internship ends:_____

Internship is: Full-time—No. hours/week:_____

                Part-time—No. hours/week:_____

Site Supervisor's Name (print):_____

Number of credits for proposed internship:_____

Credit for the internship will apply as: major credit   minor credit   elective credit

Method of Grading: pass/fail   letter grade

Total number of internship credits completed prior to this proposal:_____

Faculty Sponsor's Name:_____ Phone No.:_____

(See reverse side)

Dean's Office distributes copy of first page to Registrar's Office of registration processing   VPAA – 11/06

ATTACH RESPONSES TO SECTIONS II, III, AND IV

II.   INTERNSHIP POSITION DESCRIPTION
     a.   Location of the placement — name, address, phone.
     b.   What are the specific duties entailed in your internship? Include day-to-day tasks and specific projects, reports, attendance at conferences, special meetings, etc.

III.   EDUCATIONAL GOALS AND OBJECTIVES
     a.   Why is the internship being undertaken?
     b.   What are your specific goals and objectives?
     c.   What are your expected learning outcomes in terms of the application of theory or a method of inquiry, acquisition of professional knowledge, or development of specific skills, etc.?

IV.   METHODS OF EVALUATION
     a.   How will your agency supervisor evaluate your performance? How often and in what format?
     b.   Will the agency supervisor's evaluation be used in determining a final grade? If so, how?
     c.   What evidence can be provided to demonstrate that your educational goals and objectives have been achieved?
     d.   What evaluation data will your faculty sponsor require? For example:
        - research paper(s): documentary, analytic, creative
        - journal or log of activities, thoughts, impressions, analysis
        - written reports, essays
        - reports, papers, materials written for the agency
        - written or oral expression
        - reading lists
        - other (explain)
     e.   What contacts will you have with your faculty sponsor (on-site visit(s) by your faculty sponsor or consultation in person, by phone or mail) and how often?
     f.   Interns are required to do a written evaluation of the placement site. Give this evaluation to the faculty sponsor at the end of the internship period.

V.   APPROVAL FOR INTERNSHIP CREDIT

Signatures must be obtained in the following order:

YES   NO

               Faculty Sponsor:_____ Date:_____

               Academic Advisor:_____ Date:_____

               Dept. Chairperson:_____ Date:_____
               (of faculty granting credit)

               Site Supervisor's Signature:_____ Date:_____

               Dean:_____ Date:_____

GUIDELINES FOR INTERNSHIPS AND COLLEGE-SPONSORED EXPERIENTIAL EDUCATION*
SUNY Plattsburgh

REQUIREMENTS
1. A student receiving credit for an internship must have a faculty sponsor and an agency supervisor for the internship.
2. Interns are required to work in the internship placement for at least three hours per week (15 weeks) for each academic credit hour. Internships may be full or part-time, paid or unpaid.
3. The specific field duties and responsibilities of the intern will be agreed to by the faculty sponsor, the agency supervisor, and the intern, and they will be outlined in the learning contract.
4. Specific academic requirements appropriate to the placement will be established by the faculty sponsor and specified in the Learning Contract.

ADMISSIONS STANDARDS
1. Students wishing an internship for credit must have junior, senior, or graduate class standing. The minimum GPA for participation in an internship shall be established by the sponsoring department. Departments may establish prerequisite courses and minimum GPA in these courses.

INTERNSHIP CREDIT
1. Course Credit — Interns will be enrolled under a department course: Dept. 498 (or Dept. 598) Internship — Title. Students may take internships for major, minor, concentration, or elective credit Academic credit may range from one credit hour to a maximum of 15; any departmental policies on the maximum or minimum number of internship credits used to meet major, minor, or concentration requirements apply. Failure to complete all requirements of the internship will lead to an incomplete (policy pertaining to "I" grades will apply), or an unsatisfactory/failing grade will be given by the faculty sponsor.
2. Credit toward the baccalaureate: Usually, the total number of credits awarded for internships may not exceed 18 toward the fulfillment of the credit hours required for a bachelor's degree. The suggested limit of 18 hours does not include any credit taken by interns in related courses such as a seminar that may be taken in tandem with the internship. More than 18 credit hours of internship may be earned if the hours beyond the limit are in addition to the credit hours required for graduation.
3. Credit Hours —No more than one credit hour can be awarded for each 45 hours of internship work, 3 hours per week x 15 weeks.

QUESTIONS TO BE CONSIDERED WHEN APPROVING INTERNSHIP REQUESTS
1. Does the student have the capacity (intellectual level of maturity, independent work habits, etc.) to undertake an internship?
2. Does the student possess sufficient background in terms of course work to successfully pursue this internship?
3. Does the student have the required grade point average?
4. Does the placement provide the student adequate opportunity to achieve the proposed objectives?

*Based upon Faculty Senate Policy, May 1983.

Appendix 4

Internship Learning Contract, Center for Community Engagement, Wartburg University

Internship Learning Contract

Center for Community Engagement • (319) 352-8698 • E-mail: internships@wartburg.edu • 100 Wartburg Blvd. Waverly, IA 50677 • www.wartburg.edu/cce/internships

A. Information to be completed by Student Intern

Student Intern _____ ID# _____
　　　　　　　　　Last name　　　　　　　　　　First name

Permanent Address_____City_____State_____Zip_____ Tel (____)

E-Mail_____ Major_____CUM GPA_____ Major GPA*_____

Faculty Internship Sponsor_____Title of Internship Position_____

Course Number_____ Number of Credits_____ (Minimum 140 hours per 1 credit) Academic Year_____

Start Date_____ End Date_____ Hours Per Week_____ Number of weeks_____ Total Hours_____

Term ☐ Fall ☐ Winter ☐ May ☐ Summer　　**Year** ☐ Third ☐ Fourth　Completed Form Due:_____

B. Academic Component Description—to be completed by Student Intern and Faculty Sponsor

Learning Objectives (Attach Additional Sheets As Needed)
What do you (the student) intend to learn through your internship? List specific learning objectives in the following areas:

1. Academic knowledge (issues, subject
areas):_____

_____

2. Career-related skill
areas/experience:_____

_____

3. Integration of personal, academic, and career
issues:_____

_____

Methods of Evaluation: How do you (the student) intend to meet your learning objectives?
☐ Term Paper ☐ Weekly Log/Journal ☐ Project ☐ Portfolio ☐Presentation ☐ Other

_____

*Major GPA Calculator Can Be Accessed on the Wartburg Registrar Web Page
http://www.wartburg.edu/academics/registrar/

Appendix 5

Guidelines for Field Experience/Teaching Practicum/Internship Learning Contract, University of Northern Colorado

UNIVERSITY *of*
NORTHERN COLORADO

Higher Education &

Ph.D. **HESAL**

Student Affairs Leadership

Guidelines for Field Experience/Teaching Practicum/Internship Learning Contract

There are three different types of field experiences offered in the program in Higher Education and Student Affairs Leadership. HESA 661 offers opportunities for supervised experiences in teaching courses having student development as a primary focus. The internship, HESA 670, offers an in-depth experience in which the student gains broad exposure to the responsibilities of a particular leadership position. HESA 675 provides the opportunity for students to engage in a more focused supervised experience, usually involving working on one or more projects.

Catalog Course Descriptions:

HESA 661, Practicum in College Teaching for Student Development. Actual classroom experience in teaching any course that has developmental content as part of its objectives, while under supervision. 2 credits. S/U graded. Consent of instructor.

HESA 670, Internship in Higher Education and Student Affairs Leadership. Minimum of 18 hours per week in practical, field-based, skill-building, experiential training throughout a 16-week semester. Provides in-depth experience with leadership and/or student services delivered at the site. 6 credits, repeatable to a maximum of 18 credits. S/U graded. Consent of instructor.

HESA 675, Field Experiences in Higher Education and Student Affairs Leadership, carries a subtitle descriptive of the site in which the experience is conducted. Four hours per week per credit hour of direct involvement throughout a 16-week semester. 1–3 credits, repeatable with different subtitles, to a maximum of 12 credits. Experiential training in a field experience setting provides an overview of student service related to understanding of Higher Education and Student Affairs Leadership. S/U graded. Consent of instructor.

Guidelines and Expectations for Internships and Field Experiences:

1. Philosophy of the agency or service. The student should gain a thorough understanding of the philosophical basis for the existence of the agency/service, how it fits into the overall setting, its role and function within the larger organization, and the goals of objectives of the agency/service. Students should become knowledgeable about the principles and policies governing the operation of the agency/service.

2. Functions of the agency/service. The student should gain a working knowledge of the various functions of the agency/service, the programs and services it provides, how these are administered and funded, and the staffing patterns which exist. Insofar as possible, the student should become able to perform or oversee many of the functions and roles of the agency's staff. Where it is not possible for the student to learn an operation or provide a specific service, there should still be an effort to familiarize the student with that aspect of the operation.

3. Expectations:
   A. A Learning Contract is to be completed and signed by the student, field site supervisor, and the HESAL faculty supervisor no later than the first week of the term. See separate document entitled "HESAL Internship/Field Experience Learning Contract."
   B. Time commitments:
      HESA 661: actual class time, preparation time for each class, time for evaluation of student work, office hours, and consultation time as required by the HESAL faculty supervisor.
      HESA 670: 18–20 hours per week for the 16-week semester, plus supervisory conferences with the HESAL program faculty supervisor.
      HESA 675: 4 hours per week per credit hour for the 16-week semester, plus supervisory conferences with the HESAL program faculty supervisor.
   C. Field Site Supervisor responsibility: Provide space for the student, supply orientation and general supervision, guide the student's work while also permitting the student the freedom to explore and set personal goals and priorities. Negotiate, with the student and the HESAL program faculty supervisor, a formal learning contract. Structure and supervise student experiences. Communicate with the HESAL program faculty supervisor regarding any concerns about the student's performance. Provide a final evaluation of the student's work to the program faculty supervisor.
   D. Special projects: May be arranged on an individual basis, but should not take the place of the generalized learning expected of the student.
   E. Evaluation: The site supervisor will provide an evaluation of the performance of the student, based primarily on the manner in which the student has fulfilled the expectations of the learning contract.
   F. Ethical Standards: The student is expected to maintain ethical behavior consistent with the standards of the profession, with the requirements of the agency/service/office, and the institution. The student is expected to be familiar with, and be committed to behavior consistent with, all applicable ethical standards and laws pertaining to the field experience setting.
   G. Mutual Benefits. It is expected that there will be mutual benefit to both the student and the agency/service/office of the field experience, and care should be exercised to insure that neither the student nor the field experience site are placed in a position in which mutual benefit is jeopardized.

                                                                    revised 4/04

Appendix 6

Career Center Internship Program, California State University, Long Beach

INTERNSHIP PROGRAM

The Internship Program is designed to allow undergraduates to integrate classroom study with supervised experience in their field. The Internship Program operates as a partnership between the University and organizations from government, industry, business, and the non-profit sector.

To post an internship opportunity use **BeachLINK- Online Job & Internship Posting Service**.

- **Benefits to Employers**
- **To become an Internship Employer**
- **Unpaid Internships**

Benefits to Employers:

- Internships provide a continual pool of pre-screened, highly motivated employees.
- Internships allow you to handle staffing needs with creativity and flexibility-freeing permanent employees to do more advanced or higher priority work.
- Internships provide completion of special projects and fill the gap during peak work loads.
- Internships reduce recruitment costs and lower training costs.
- New ideas and fresh insights are brought to the company.
- Internship students returning to campus are great public relations agents-they have a very positive effect for future recruiting and hiring efforts.
- Internships help maintain an on-going positive relationship with the university that complements the organizational or corporate goals that focus on community involvement.

To become an Internship Employer:

- Determine which functional area would be best served by using students from the program and develop written position descriptions.
- Determine who will select, supervise, mentor and evaluate students.
- Develop an orientation or training plan for students.
- Review **Internship Program Guidelines** according to the CSULB Employer Services Guidelines for approval guidelines.
- Third Party Recruiters please refer to **Services for Third Party Recruiter Guidelines** according to the CSULB Employer Services Guidelines.
- To post an internship opportunity use **BeachLINK- Online Job & Internship Posting Service**

Unpaid Internships:

The U.S. Department of Labor established the following 6 criteria in the **Fair Labor Standards Act** (FLSA) to determine if an internship or training program can be excluded from the minimum wage requirement.

1. The internship, even though it includes actual operation of the facilities of the employer, is similar to training which would be given in an educational environment;
2. The internship experience is for the benefit of the intern;
3. The intern does not displace regular employees, but works under close supervision of existing staff;
4. The employer that provides the training derives no immediate advantage from the activities of the intern; and on occasion its operations may actually be impeded;
5. The intern is not necessarily entitled to a job at the conclusion of the internship; and
6. The employer and the intern understand that the intern is not entitled to wages for the time spent in the internship.

(Note: The FLSA makes a special exemption for individuals who volunteer for a state or local government agency and for individuals who volunteer for humanitarian, religious, charitable, civic purposes to non-profit organizations.)

If you have any questions, please **contact the Internship Program Coordinator** at 562-985-5552.

California State University, Long Beach
1250 Bellflower Boulevard, Long Beach, California 90840

References

Distance Learning Information Packet. (2011). Syracuse University, Syracuse, NY (Appendix 8).

Guidelines for Field Experience/Teaching Practicum/Internship Learning Contract. (2011). University of Northern Colorado, CO (Appendix 5).

Internship Learning Contract, *Gallatin school of individualized study*. New York, NY: New York University (Appendix 1).

Internship Learning Contract. (2011). Warturg College, Center for Community Engagement, Waverly, IA (Appendix 4).

Internship Learning Contract Tips. (2010–2011). University of Minnesota, Minneapolis, MN (Appendix 2).

Internship Program Brochure, *Career development center*. Long Beach, CA: CSU Long Beach (Appendix 6).

Learning Contract and Internship Proposal. (2011). SUNY Plattsburgh, Plattsburgh, NY (Appendix 3).

Principles of Professional Practice for Career Service and Employment Professionals. (2011). National Association of Colleges and Employees (NACE), Bethlehem, PA (Appendix 7).

10 Evaluation of Internships

This chapter is concerned with the evaluation of the performance of individual interns. The evaluation of an individual intern's performance will involve consideration of several components (Buckingham, 2007):

1. *Professional Issues*:
   The professional issues involved in the intern's performance will include:
   I. *Observance of work hours*: If the intern is required to sign in and out of the workplace, this section of the performance evaluation may include a numerical report of the number of late days or—alternatively—whether the intern has met every required standard set at the beginning of the internship; for example, not more than 5% late attendance.
   II. *Meeting deadlines and effective use of time*: This section will note the number of times the intern has been late in submitting any reports or mini-reports or work assignments.
   III. *Dress and appearance (Darlington and Schuman, 2008)*: The intern should dress and groom himself or herself appropriately to the organization and/or event(s) within the organization. With a larger number of first generation and historically underrepresented minority groups going for careers in science and engineering than ever before, a firm or company that employs interns may wish to have workshops on professional issues for these students.

2. *Organizational Issues*:
   I. *Organization's mission, vision, and goals*: Understanding the organization's mission, vision, and goals is important for the success of the intern. This knowledge can come through any formal component of a training program by a human resources department and can include lectures or a webinar on these topics, preferably given by a middle- to upper-level manager.
   II. *Operates within the normal expectations of the organization*: The norms may be unspoken ones, and one should not be too harsh on the intern who may be present in the organization for a limited amount of time.
   III. *Observing the chain of command; confidentiality issues*: Since the typical intern might be experiencing the workplace for the first time through the internship, the human resources department might see fit to have formal lectures, especially on any confidentiality issues. While it is desirable not to have the intern become exposed to confidential information and/or issues, it may also be unavoidable in the context of modern competitive industry.
   IV. *Career Issues (Bolles, 2011)*:
      a. Does the intern do a good self-evaluation, and does he or she understand his or her own strengths and weaknesses?
      b. Has the intern shown reasoned judgment relating to choosing a future career?
   V. *Communication and Teamwork (Johnson, 1978; Upcraft and Schuh, 1996)*:
      a. How does the intern relate to coworkers?

Lifelong Learning for Engineers and Scientists in the Information Age. DOI: 10.1016/B978-0-12-385214-4.00010-6

    b. How does the intern resolve conflicts?

    c. How are the intern's written communication skills?

        1. Grammar?

        2. Spelling?

        3. Communication of concepts?

        4. Communication of mathematical ideas, concepts, and thoughts? Are they simplified, yet accurate?

    d. How are the intern's oral communication skills?

        1. Listening skills?

        2. Following verbal directions?

        3. Participation in meetings?

        4. Overall effectiveness in oral communication?

VI. *Technical Competence*:

    a. Does the intern have a basic understanding of the subject matter?

    b. Does the intern read materials (papers, handouts, books, journals, catalogs) to keep pace with the latest developments in the field?

    c. Does the intern show adequate, inadequate, or superior performance on the job in both the theoretical and the hands-on (with equipment, instruments) aspects of the job?

    d. How is the quality of the intern's written reports (technical accuracy)?

VII. *Lifelong-Learning Skills*:

    a. Does the intern attend seminars and conferences beyond the call of duty?

    b. Is the intern showing inquisitiveness about new ideas, thoughts, or subject matters?

    c. Does the intern learn from his or her and from coworkers' experiences?

References

Bolles, R. N. (2011). *What color is your parachute: A practical guide for job-hunters and career-changers* (40th Anniversary Edition). Ten Speed Press.

Buckingham, M. (2007). *Go put your strengths to work*. New York, NY: Free Press.

Darlington, J., & Schuman, N. (2008). *The everything job interview book*. Avon, MA: Adams Media.

Johnson, D. W. (1978). *Human relations and your career*. Englewood Cliffs, NJ: Prentice Hall.

Upcraft, M. L., & Schuh, J. H. (1996). *Assessment in student affairs*. San Francisco, CA: Jossey-Bass.

11 Information Literacy and Career Skills

Introduction

Generally speaking, academic librarians do not offer workshops on career planning, leaving it to their colleagues at the career center. Over the past 20 years, academic librarians have concentrated on developing a solid and sound foundation collaborating with teaching faculty in developing and embedding information literacy (IL) assignments. Librarians teach students IL skills such as finding references (citations) on a topic: this has been covered in detail in other chapters of this book. Traditional information research concentrates effort on finding authoritative peer-reviewed articles and evaluates them for relevance for the research task.

Librarians are more recently turning their attention to the cocurricular units on campus. Progress has been made in such areas as international student programs. The same information-seeking skills used for research papers can be used for career planning and enhancing the search for internships and jobs. In this case, one focuses on finding non-peer-reviewed formats such as trade magazines, newspapers, newswires, and company and industry data in order to gain an understanding of the chosen field. Developing career skills are a vital part of lifelong learning, and IL provides an essential part in that skill set.

Librarians and career counselors possess complementary expertise and make good partners in co-creating programs to increase the lifelong-learning skills of their students. In passing we should note that public libraries have an excellent record in providing career information and their programs have sadly grown with the downturn in the economy.

Collaboration Between Career Counselors and Librarians

Career centers are considered to be part of the cocurricular programs, together with other student services such as international student services, residential life groups, writing centers and learning centers, whereas libraries are usually part of academic divisions and primarily support curricular and research activities. This administrative division may have lead to the traditional lack of collaboration between career counselors and librarians. Abel (1992) reported on a survey that she sent out

Lifelong Learning for Engineers and Scientists in the Information Age. DOI: 10.1016/B978-0-12-385214-4.00011-8

to 50 colleges in the Midwest to a mix of libraries and career planning offices on campuses. She summarizes the findings:

> *In essence, there does not seem to be a great deal of sharing of information between career planning departments and libraries. While excellent programs and supports are available at many institutions such as Harvard's Baker Library and the University of Illinois for example, institutions with fewer resources and need for resource sharing have not take the cue from the larger, better funded institutions. Efforts to provide access to each other's resources seem rather lackluster. (p. 56)*

If a similar survey is sent out today, there is some evidence to suggest that the situation is slowly changing. Abel (1992) continues to describe the excellent model of the Career Information Center developed at DePaul University, which is staffed by a reference librarian. The library works in tandem with the college's Career Planning Information Center (CP & P). The librarian researches recruiters and provides info to the CP & P staff, gives tours of the library, and gives seminars to the CP & P staff. Abel's article is part of a special issue of *Reference Librarian* edited by Byron Anderson. The topic was further explored in another issue of the *Reference Librarian* edited by Elizabeth Lorenzen (1996) to bring it up-to-date with the advent of the Internet ("Career Planning and Job Searching in the Information Age").

Happily there is a growing awareness of the advantages of career counselors and librarians blending their respective expertise to reach the common goals of student success and the pursuit of lifelong learning. A literature review revealed some excellent programs of collaborations between career centers and university libraries: DeHart (1996), Dugan, Bergstrom, and Doan (2009), Hollister (2005), Love (2009) and Song (2005). Brian DeHart expresses it succinctly and suggests a win–win situation for all the concerned parties:

> *Collaboration between the academic library and the university placement office is a natural partnership. Joint planning and consultation in both collection development and programming evolve into more ambitious endeavors such as instruction. At DePaul University, this relationship has become well established through the efforts of library administration and reference/instruction librarians. The combined effort has resulted in students' increased use of library resources in order to prepare better for the job market, while gaining life-long skills for finding and analyzing career information.*

The rest of this chapter is devoted to the model developed at California State University, Long Beach (CSULB), as an example of how these two sets of professionals can combine their skills to assist students in uncovering internships and job opportunities.

Locating Internships

The best place to start a search for internships is at the university or college's career center. Most career centers have an established program of networking and maintaining links with local companies and industries. Employers advertise their internships via the career center. CSULB's Career Development Center (CDC) is a good example. "BeachLINK," is a local database maintained by the CSULB Career Center for students and "on average there are over 400 internship postings on BeachLINK daily" in all fields (not just engineering) and "these internships have all been previously reviewed by the Internship Program Coordinator to meet the requirements for academic credit through a CDC Internship Course" (http:// www.careers.csulb.edu/job_search/internships/internship_resources.htm; retrieved September 19, 2011). Searching the Internet for internships is always an option, but it can be overwhelming wading through thousands of hits. It is preferable to use a metasite, portal, aggregator, or service that has already evaluated the sites for one's convenience. One good way to do it is to use the expertise of career counselors who monitor and provide information on the best Internet sites. For instance, the CSULB Career Center has an excellent web page that provides a good starting point for research: http://careers.csulb.edu/job_search/internships/internship_resources.htm. Here are a few excellent sites:

- The Wellesley College Career Center, http://www.wellesley.edu/cws/findingint.pdf.
- Florida Institute of Technology has a good site on internships for Engineering and Technology Majors, http://www.fit.edu/career/students/resources/engineering.php.
- Internship Programs.com, http://www.internshipprograms.com/, is a well-organized site that can be searched by field (e.g., engineering).
- Get that Gig, getthatgig.com, is a more modern looking site providing information on internships, but it also acts as a metasite to other sites for internships and (for instance) entry-level jobs in engineering. The engineering page has links to other useful engineering resources from associations and other sources, http://www.getthatgig.com/cat_engineering.html
- After College, aftercollege.com, connects college students, alumni, and employers through faculty and career networks at colleges and universities across the country.
- Internweb, http://www.internweb.com/, allows employers to post to their site without a charge and also gives tips to employers on designing internship programs.
- JobWeb, http://www.jobweb.com/students.aspx?folderid = 86, offers career and job-search advice for new college graduates, and is the online complement to the Job Choices job-search publications.

Knowledge of professional associations and their resources is a very important aspect of lifelong learning. Most professional associations offer student membership at a much reduced rate. Many of the associations provide valuable career resources for their student members and these should be utilized in addition to the resources from the career center and the Internet. A few examples are mentioned:

- Institute of Electrical and Electronics Engineering (IEEE), http://www.ieeeusa.org/careers/ student.menu.html
- Association of Computing Machinery (ACM), http://jobs.acm.org/home/index.cfm?site_ id=1603
- American Society of Civil Engineers (ASCE), http://careers.asce.org/jobs
- American Society of Mechanical Engineers (ASME), http://jobboard.asme.org/jobs
- American Institute of Chemical Engineers (AIChE), http://careerengineer.aiche.org/c/ search.cfm?site_id=1932

Finding Background Information on Companies

Once some internship opportunities have been identified (using the resources mentioned above), it is a good idea to do some research on the company. Some of the databases via the campus library are very useful for this task. The information found in journals—especially trade magazines, articles, and newspapers—can be used to write targeted cover letters citing (for instance) the company's current projects or recently awarded contracts. The same databases can be used to prepare for interviews or for meeting company representatives at job fairs. In general, the databases in the fields of business and management offer the best choices for research on company information. *ABI Inform*, for instance, is one of the most comprehensive and well-known databases in business, with over 3,200 journals (over 2,460 full-text titles) covering business and economic conditions, corporate strategies, management techniques, and competitive and product information. Its international coverage gives researchers a complete picture of companies and business trends around the world. Most 4-year campuses have access to ABI Inform or other similar databases. *Hoovers.com*, another major resource, is a respected company of business analysts who provide authoritative information on big public companies. Hoovers provides some free information on their web site (hoovers.com), but more information is available through such databases as ABI Inform. Hoovers gives you contact information, names of key players, financial information, and names of competitors (very important).

Engineering and science databases (such as *Compendex* and *Web of Science*) can also be used in an indirect way to search for information about larger companies with active research and development programs whose staff actively publish in the professional and research journals. For the enthusiastic student, searching for patents by the company may also offer interesting insights on their intellectual property. Google Patents offers the easiest way to search for US Patents, http://www.google.com/patents.

Internet Versus Library Databases

One may ask: why not simply use the Internet to search for information about companies and industries? This is a question often asked by students and others. The Internet is a very inefficient way to search for targeted information. Even given the sophistication reached by many search engines, such as Google, too many hits are retrieved, and it is so hard to determine or evaluate the relevant hits that the searcher feels overwhelmed. And, more importantly, one can only access the free information, whereas the subscription databases give us access to proprietary information that is not freely available on the Internet. On the Internet, users have to pay exorbitant costs online to access articles individually. The databases available via the academic library are updated constantly—often on a daily basis. Also, the databases offer us value-added tools in terms of fielded searching, which help us focus our search with increased precision and efficiency. For instance, we can search (to name two examples) by geographical locations and NAICS numbers. North American Industry Classification System (NAICS) is the standard used by Federal statistical agencies in classifying business establishments for the purpose of collecting, analyzing, and publishing statistical data related to the US business economy (NAICS replaced Standard Industrial Classification (SIC) codes), http://www.census.gov/eos/www/naics/. Most, if not all, business databases use NAICS numbers as a way of categorizing information about companies and businesses.

The CSULB Model: Job Seeking in Economically Hard Times

The economic downturn is highlighting the need for students to use innovative ways to "create" their own internships and jobs. "Business Week" has an excellent video with tips on how to land a great internship in today's tough economy (http://www.businessweek.com/managing/content/dec2008/ca2008122_876870. htm?chan=careers_managing+index+page_top+stories; retrieved August 14, 2011).

Given this backdrop, the CSULB engineering librarian and a career counselor developed a workshop for engineering majors to address the current situation. The theme of the drop-in workshop is searching for "hidden jobs"—which can be applied equally to internships. For our purposes, we defined "hidden jobs" as jobs (or internships) that may be "hidden" and not advertised via the usual channels. Or perhaps if a suitable CV and cover letter lands on an engineer's desk, they may decide to create an internship where none existed.

The workshop teaches students to use career center and library resources seamlessly to find potential employers in engineering. Once they have identified some companies, students can draft a "prospecting letter" asking about possible internships (and jobs). The prospecting letter has the following elements:

- Indicates interest and reveals the source of information,
- Outlines qualifications and describes how these qualifications match the work environment,
- Suggests an action plan,
- Requests an interview, and
- Expresses appreciation to the reader for his or her time and consideration.

The letter is enhanced by citing some interesting (very brief) up-to-date information about the company's current project(s). Research in the library databases may also reveal the award of a major contract indicating that the company will be hiring many new employees in the next few months—and the student can get ahead of the pack by sending out the letter and CV right away.

Example of a prospecting letter:

Dear Ms Diaz,
I read about Fluor Corp. in *Engineering News Record* and decided to research your company. I am very interested in working for you. I will receive my BS in Mechanical Engineering this May from CSULB. While at CSULB, I was heavily involved in the local chapter of ASME and was the Treasurer from 2008–2009.

As you will see from my resume, I have taken many classes that are in areas that your company is involved in (*name a few classes*). I am very interested in working for a global company such as yours. I have traveled extensively in the Far East (Japan, China, and Singapore to name a few countries) and am interested in working on international projects. I was particularly interested to hear that Fluor announced recently it had been awarded the Project Hijau Gasoil Phase-1 by the Shell Refining Company, Malaysia. (Real Estate & Investment Business, October 16, 2010).
Etc.
Etc.

Undergraduate engineering students are the target audience for this workshop. Most undergraduates have not generally been exposed to company and industry information. They may have heard about one or two big local companies (in the case of CSULB, it is Boeing and Northrop Grumman) but are generally not aware of the range and array of companies. It is a different case with graduate students, especially at such a big state institution as CSULB where many have worked in industry for awhile or are working part-time and are more aware of the engineering field. International students have a different set of needs, often searching for companies in their home countries.

The library portion of the workshop focuses on the business databases available via the library. Most academic libraries will have access to at least one or two excellent business databases. The following topics are covered:

- Finding possible companies in an industry segment,
- Finding useful information about those companies,
- Information about competitors (which identifies similar companies),
- Secondary information about companies from trade magazines, newspapers, and business analysts, and
- Finding companies within an industry and geographical location.

It should be emphasized that the searches do *not* give information on vacancies or internships, but give a good "snapshot" on a variety of companies and industries. Students can follow up by going to the company's home page on the Internet to ascertain if there are any vacancies or internships or by simply sending a "prospecting" letter citing some information about the company they found in their research. Given the competitive nature of the industry, it is hoped that these letters will stand out amongst other applicants and indicate to potential employers that the applicant has good information-seeking (lifelong-learning) skills and has taken the time to research their company.

Without a doubt, this type of research catches the students' attention much more than searching databases to find references for research papers. Students immediately see the connection between developing good information-seeking skills and landing an internship and eventually a job. As mentioned above, but worth emphasizing, developing IL skills to search for internships and jobs is an essential part of lifelong learning. These skills are not just important during an academic career and at graduation, but also throughout the working life of an engineer. It is a fact that engineers, especially engineers of the future, will have to change jobs a few times in their careers given the fast-paced world of technology and a dynamic economy. Knowing how to research career prospects is an important skill and an essential part of lifelong learning.

An added benefit of a workshop such as this is that it introduces students—especially undergraduate students—to a broad array of business resources. The skills they develop can be applied to other courses in their program. For instance, they can apply these skills to design courses (especially capstone design courses), which require them to find product and market data for their design or invention. Ultimately, it may also spark an interest in entrepreneurship.

Assignment Combining Career and Library Skills

One of the spin-offs of the CSULB career center collaboration was the development of an assignment that was incorporated into the syllabus of a freshmen engineering course.

Students were required to complete the following tasks:

- Create a resume,
- Pick an industry of interest (e.g., robotics, computer animation, solar energy, and so on),
- Use ABI Inform to find companies in that industry,
- Research one of the companies,
- Write a prospecting letter to the company citing one or two of their recent projects, and
- Make a brief presentation at the end of the semester on their company.

By the end of the semester, each student increased his or her knowledge about engineering companies "n" times—"n" being the number of students in the class.

In one of the sections, the instructor decided to mail off the letters and resumes—with the students' consent—and it was reported that one student had been offered an internship even before the end of the semester.

Sustaining the Collaboration

There is a major workload issue on academic campuses everywhere; librarians are already hard-pressed to keep up with supporting IL programs in their respective curricular programs in addition to all of their other duties. One way forward is to follow the "train the trainer" model. Librarians should offer training sessions to the career counselors on the basics of searching the library's databases so the counselors can incorporate these skills seamlessly into their own workshops and daily inter-actions with the students. This may be the most sustainable way to move forward, thereby also enhancing the information skills of the counselors. This evolution has already started to happen in an organic way at CSULB.

References

Abel, C. (1992). A survey of cooperative activities between career planning departments and academic libraries. *The Reference Librarian, 36*, 51.

DeHart, B. (1996). Job search strategies: Library instruction collaborates with university career services. *The Reference Librarian, 55*, 73–81.

Dugan, M., Bergstrom, G., & Doan, T. (2009). Campus career collaboration: "Do the research. Land the job." *College & Undergraduate Libraries, 16*(2), 122–137. 10.1080/10691310902958517.

Hollister, C. (2005). Bringing information literacy to career services. *Reference Services Review, 33*(1), 104–111. 10.1108/00907320510581414.

Lorenzen, E. A. (1996). *Career planning and job searching in the information age.* New York: Haworth Press. (Special issue of the Reference Librarian , No. 55).

Love, E. (2009). A simple step: Integrating library reference and instruction into previously established academic programs for minority students. *Reference Librarian, 50*(1), 4–13. 10.1080/02763870802546357.

Song, Y. (2005). Collaboration with the business career services office: A case study at the University of Illinois at Urbana-Champaign. *Research Strategies, 20*(4), 311–321.

12 Conclusion

The material presented in this work is intended to be a guide and a toolkit for those engaged in the education and professional development of engineering students and engineers making their way along a career path. It is also hoped that some of the models presented in this text will inspire the development of an even greater variety of practices that can be used in the classroom and in the workplace. In this digital age of electronic information—with the constant flow of raw data and unvetted topics labeled "important"—it is critical that best practices flourish with the unified guidance of ABET, Inc. (formerly the Accreditation Board of Engineering and Technology), the Association of College and Research Libraries (ACRL), and the host of engineering educators and practitioners who daily put into action the highest objectives for educating engineers.

Developing adaptable engineers with critical-thinking skills—people who can evolve with the times—will require the updating and sharing of information. The references cited and discussed in this book should lead to additional writings. The classroom tools and assessment techniques should inspire new ways of teaching. These successful classroom interactions and expositions should then be discussed at conferences and workshops. These findings should, in turn, inform and influence committees at ABET and ACRL to update and refine information literacy and lifelong-learning standards. Just as the technologies that engineers create and work with are developed into newer and more dynamic processes and components, so too must educators and employers evolve.

Lifelong Learning for Engineers and Scientists in the Information Age. DOI: 10.1016/B978-0-12-385214-4.00012-X

Printed in the United States
By Bookmasters